ENVIRONMENTAL SCIENCE, ENGINEERING AND TECHNOLOGY

U.S. WEATHER SATELLITES

BACKGROUND, PROGRAM CHALLENGES AND POTENTIAL DATA GAPS

ENVIRONMENTAL SCIENCE, ENGINEERING AND TECHNOLOGY

Additional books in this series can be found on Nova's website under the Series tab.

Additional e-books in this series can be found on Nova's website under the e-book tab.

ENVIRONMENTAL SCIENCE, ENGINEERING AND TECHNOLOGY

U.S. WEATHER SATELLITES

BACKGROUND, PROGRAM CHALLENGES AND POTENTIAL DATA GAPS

LOUISE O. BERKAN
EDITOR

nova publishers
New York

Library of Congress Cataloging-in-Publication Data

ISBN: 978-1-63482-726-3

Published by Nova Science Publishers, Inc. † New York

CONTENTS

PREFACE

This book assesses progress on program schedule, cost, and functionality; assesses efforts to identify and address issues discovered during integration and testing; and evaluates the likelihood of a gap in satellite coverage and actions to prevent or mitigate such a gap. This book also evaluates NOAA's progress on the Joint Polar Satellite System (JPSS) satellite program with respect to cost, schedule, and mitigation of key risks; identifies the benefits and challenges of alternatives for polar satellite gap mitigation; and assesses NOAA's efforts to establish and implement a comprehensive contingency plan for potential gaps in polar satellite data.

Chapter 1 – NOAA, with the aid of the National Aeronautics and Space Administration (NASA), is procuring the next generation of geostationary weather satellites. The GOES-R series is to replace the current series of satellites, which will likely begin to reach the end of their useful lives in 2015. This new series is considered critical to the United States' ability to maintain the continuity of satellite data required for weather forecasting through 2036. GAO was asked to evaluate GOES-R. GAO's objectives were to (1) assess progress on program schedule, cost, and functionality; (2) assess efforts to identify and address issues discovered during integration and testing; and (3) evaluate the likelihood of a gap in satellite coverage and actions to prevent or mitigate such a gap. To do so, GAO analyzed program and contractor data, earned value data information, and defect reports, compared both defect management policies and contingency plans to best practices by leading organizations, and interviewed officials at NOAA and NASA.

Chapter 2 – NOAA established the JPSS program in 2010 to replace aging polar satellites and provide critical environmental data used in forecasting weather and measuring variations in climate. However, GAO and NOAA have previously reported that a gap in satellite data between the current satellite and the next one is likely. Given the criticality of satellite data to weather forecasting, the likelihood of a significant satellite data gap, and the potential impact of a gap on the health and safety of the U.S. population and economy, GAO added this issue to its High Risk List in 2013. GAO was asked to review the JPSS program. GAO's objectives were to (1) evaluate NOAA's progress on the JPSS satellite program with respect to cost, schedule, and mitigation of key risks; (2) identify the benefits and challenges of alternatives for polar satellite gap mitigation; and (3) assess NOAA's efforts to establish and implement a comprehensive contingency plan for potential gaps in polar satellite data. To do so, GAO analyzed program management status reports, milestone reviews, and risk data; examined polar gap contingency plans; and interviewed experts as well as agency and contractor officials.

In: U.S. Weather Satellites
Editor: Louise O. Berkan
ISBN: 978-1-63482-726-3

Chapter 1

GEOSTATIONARY WEATHER SATELLITES: LAUNCH DATE NEARS, BUT REMAINING SCHEDULE RISKS NEED TO BE ADDRESSED*

United States Government Accountability Office

WHY GAO DID THIS STUDY

NOAA, with the aid of the National Aeronautics and Space Administration (NASA), is procuring the next generation of geostationary weather satellites. The GOES-R series is to replace the current series of satellites, which will likely begin to reach the end of their useful lives in 2015. This new series is considered critical to the United States' ability to maintain the continuity of satellite data required for weather forecasting through 2036.

GAO was asked to evaluate GOES-R. GAO's objectives were to (1) assess progress on program schedule, cost, and functionality; (2) assess efforts to identify and address issues discovered during integration and testing; and (3) evaluate the likelihood of a gap in satellite coverage and actions to prevent or mitigate such a gap. To do so, GAO analyzed program and contractor data, earned value data information, and defect reports, compared both defect management policies and contingency plans to best practices by leading organizations, and interviewed officials at NOAA and NASA.

* This is an edited, reformatted and augmented version of a United States Government Accountability Office publication, No. GAO-15-60, dated December 2014.

What GAO Recommends

GAO is recommending that NOAA address shortfalls in its defect management approach, reduce the number of open high-priority defects, and add information to its satellite contingency plan. NOAA concurred with GAO's recommendations and identified steps it plans to take to implement them.

What GAO Found

The National Oceanic and Atmospheric Administration (NOAA)'s Geostationary Operational Environmental Satellite-R (GOES-R) program has made major progress in developing its first satellite, including completing testing of satellite instruments. However, the program continues to face challenges in the areas of schedule, cost, and functionality. Specifically, the program has continued to experience delays in major milestones and cost overruns on key components. Also, in order to meet the planned launch date, the program has deferred some planned functionality until after launch, and program officials acknowledge that they may defer more.

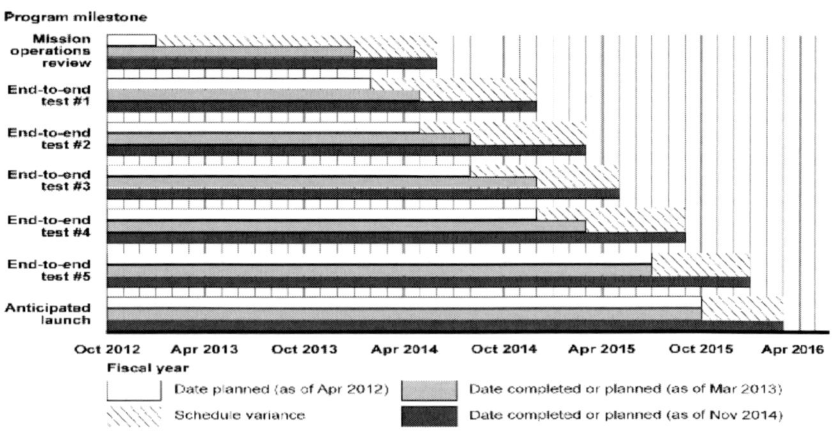

Source: GAO analysis of NOAA data. | GAO-15-60.

Delays in Key GOES-R Program Milestones.

NOAA and its contractors have implemented a defect management process as part of their overall testing approach that allows them to identify, assess, track, resolve, and report on defects. However, shortfalls remain in how defects are analyzed and reported. For example, contractors manage, track, and report defects differently due to a lack of guidance from NOAA. Without consistency among contractors, it is difficult for management to effectively prioritize and oversee defect handling. In addition, more than 800 defects in key program components remained unresolved. Until the program makes progress in addressing these defects, it may not have a complete picture of remaining issues and faces an increased risk of further delays to the GOES-R launch date. As the GOES-R program approaches its expected launch date of March 2016, it faces a potential gap of more than a year during which an on-orbit backup satellite would not be available. This means that if an operational satellite experiences a problem, there could be a gap in GOES coverage. NOAA has improved its plan to mitigate gaps in satellite coverage. However, the revised plan does not include steps for mitigating a delayed launch, or details on potential impacts and minimum performance levels should a gap occur. Until these shortfalls are addressed, NOAA management cannot fully assess all gap mitigation strategies, which in turn could hinder the ability of meteorologists to observe and report on severe weather conditions.

ABBREVIATIONS

ABI	Advanced Baseline Imager
CDR	critical design review
EVM	earned value management
FOR	flight operations review
GLM	Geostationary Lightning Mapper
GOES	Geostationary Operational Environmental Satellite
GOES-R	Geostationary Operational Environmental Satellite – R series
MDR	mission definition review
MOR	mission operations review
NASA	National Aeronautics and Space Administration
NOAA	National Oceanic and Atmospheric Administration
ORR	operational readiness review
PDR	preliminary design review
SDR	system definition review
SIR	system integration review

December 16, 2014

The Honorable Lamar S. Smith
Chairman

The Honorable Eddie Bernice Johnson
Ranking Member
Committee on Science, Space, and Technology
House of Representatives

Geostationary environmental satellites play a critical role in our nation's weather forecasting. These satellites—which are managed by the Department of Commerce's National Oceanic and Atmospheric Administration (NOAA)—provide information on atmospheric, oceanic, climatic, and solar conditions that help meteorologists observe and predict regional and local weather events. They also provide a means of identifying the large-scale evolution of severe storms, such as forecasting a hurricane's path and intensity.

NOAA, through collaboration with the National Aeronautics and Space Administration (NASA), is procuring the next generation of geostationary weather satellites, called the Geostationary Operational Environmental Satellite–R (GOES-R) series. The GOES-R series consists of four satellites and is to replace the current series of geostationary environmental satellites as they reach the end of their useful lives. This new series is expected to provide the first major improvement in the technology of GOES instruments since 1994 and, as such, is considered critical to the United States' ability to maintain the continuity of data required for weather forecasting through the year 2036. NOAA is facing a potential gap where it may not have a backup satellite in orbit between now and the launch of the first GOES-R satellite. Because of the criticality of satellite data to weather forecasting, the likelihood of data gaps in NOAA's satellite programs, and the potential impact of a gap on the health and safety of the U.S. population and economy, we added mitigating gaps in weather satellite data to our High Risk List in 2013.[1]

This report responds to your request that we review NOAA's GOES-R program. Specifically, our objectives were to (1) assess progress on the GOES-R program with respect to planned schedule, cost, and functionality; (2) assess efforts to identify and address issues discovered during integration and testing; and (3) evaluate the likelihood of a gap in satellite coverage and analyze the adequacy of contingency actions in place to prevent or mitigate such a gap.

To assess NOAA's progress in developing GOES-R with respect to cost, schedule, and functionality, we analyzed monthly program status briefings to identify current status, recent development challenges, and both expected and potential changes in functionality. We compared current and past status briefings to determine delays over time to key milestones. We also analyzed cost reserve and earned value data information to understand the program's current cost posture. To assess NOAA's efforts to identify and address issues discovered during integration and testing, we analyzed defect management policies and practices against criteria from industry best practices. We also compared defect trend data and metrics over time, and interviewed contractor officials responsible for testing and defect management. To ensure the reliability of cost and defect data, we compared data across monthly reports and other agency data sources, compared formulas from the GAO Cost Guide[2] to the program's earned value management approach, and sought corroboration from agency and contractor officials. To evaluate the likelihood of a gap in satellite coverage and analyze contingency actions, we analyzed program documentation to determine the likely length of a gap, and followed up on our previous effort to compare the GOES-R contingency plan to best practices in contingency planning identified by leading organizations.[3] We also interviewed NOAA and GOES-R program officials regarding program status, defect management, and contingency planning.

We conducted this performance audit from January 2014 to December 2014 in accordance with generally accepted government auditing standards. Those standards require that we plan and perform the audit to obtain sufficient, appropriate evidence to provide a reasonable basis for our findings and conclusions based on our audit objectives. We believe that the evidence obtained provides a reasonable basis for our findings and conclusions based on our audit objectives. See appendix I for a complete description of our objectives, scope, and methodology.

BACKGROUND

Since the 1970s, geostationary satellites have been used by the United States to provide meteorological data for weather observation, research, and forecasting. NOAA's National Environmental Satellite, Data, and Information Service is responsible for managing the civilian operational geostationary satellite system, called GOES. Geostationary satellites can maintain a constant view of the earth from a high orbit of about 22,300 miles in space.

NOAA operates GOES as a two-satellite system that is primarily focused on the United States (see fig. 1). These satellites provide timely environmental data about the earth's atmosphere, surface, cloud cover, and the space environment to meteorologists and their audiences. They also observe the development of hazardous weather, such as hurricanes and severe thunderstorms, and track their movement and intensity to reduce or avoid major losses of property and life. The ability of the satellites to provide broad, continuously updated coverage of atmospheric conditions over land and oceans is important to NOAA's weather forecasting operations.

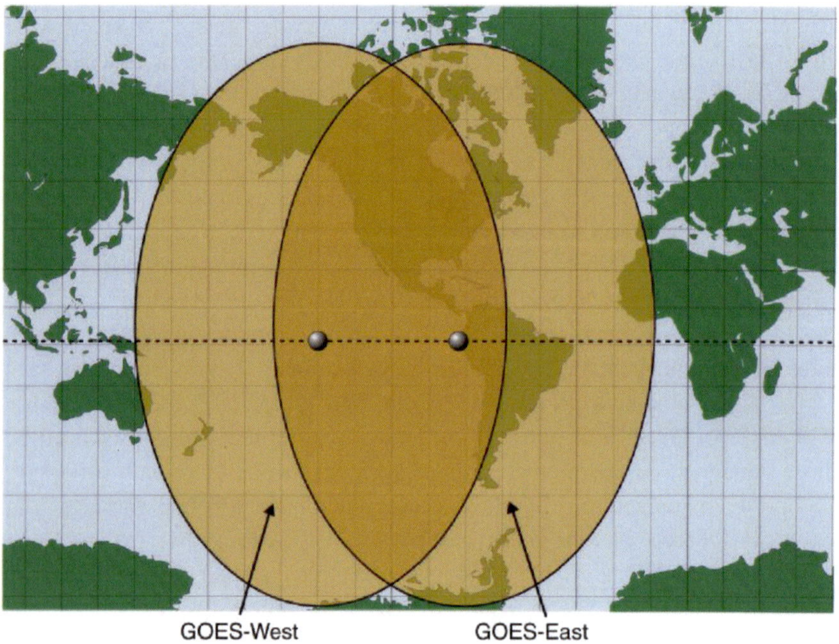

Source: NOAA (data), Mapart (map). | GAO-15-60.

Figure 1. Approximate Geographic Coverage of the Geostationary Operational Environmental Satellites.

To provide continuous satellite coverage, NOAA acquires several satellites at a time as part of a series and launches new satellites every few years (see table 1). NOAA's policy is to have two operational satellites and one backup satellite in orbit at all times.

Table 1. Summary of the Procurement History of the Geostationary Operational Environmental Satellites

Series name	Procurement duration[a]	Satellites[b]
Original GOES[c]	1970-1987	1, 2, 3, 4, 5, 6, 7
GOES I-M	1985-2001	8, 9, 10, 11, 12
GOES-N	1998-2010	13, 14, 15, Q[d]
GOES-R	2008-2024	R, S, T, U

Source: GAO analysis of NOAA data. | GAO-15-60

[a]Duration includes time from contract award to final satellite launch.

[b]Satellites in a series are identified by letters of the alphabet when they are on the ground (before launch) and by numbers once they are in orbit.

[c]The procurement of these satellites consisted of four separate contracts for (1) two early prototype satellites and GOES-1, (2) GOES-2 and -3, (3) GOES-4 through -6, and (4) GOES-G (failed on launch) and GOES-7.

[d]NOAA decided not to exercise the option for this satellite.

Three viable GOES satellites—GOES-13, GOES-14, and GOES-15—are currently in orbit.[4] Both GOES-13 and GOES-15 are operational satellites, with GOES-13 covering the eastern United States (GOES-East in Figure 1) and GOES-15 covering the western United States (GOES-West). GOES-14 is currently in an on-orbit storage mode and is available as a backup for the other two satellites should they experience any degradation in service. The GOES-R series is the next generation of satellites that NOAA is planning.

Each of the operational geostationary satellites continuously transmits raw environmental data to NOAA ground stations. The data are processed at these ground stations and transmitted back to the satellite for broadcast to primary weather services and the global research community in the United States and abroad.

Raw and processed data are also distributed to users via ground stations through other communication channels, such as dedicated private communication lines and the Internet. Figure 2 depicts a generic data relay pattern from a geostationary satellite to the ground stations and commercial terminals.

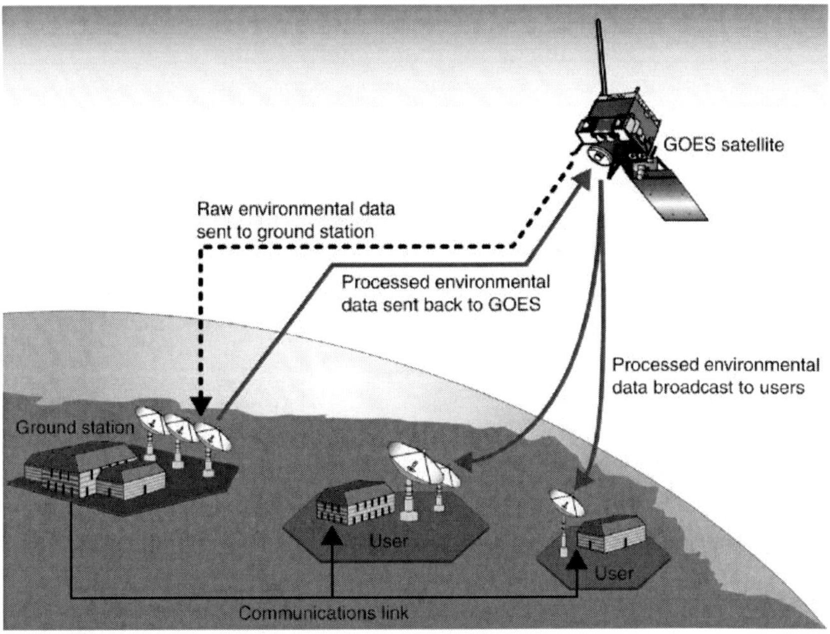

Source: GAO analysis of NOAA data. | GAO-15-60.

Figure 2. Generic Data Relay Pattern for the Geostationary Operational Environmental Satellites.

Overview of the GOES-R Program

NOAA established the GOES-R program to develop and launch the next series of geostationary satellites and to ensure the continuity of geostationary satellite observations. The GOES-R satellite series is designed to improve upon the technology of the prior satellite series in terms of system and instrument improvements. NOAA expects that the GOES-R series will significantly increase the clarity and precision of the observed environmental data. In addition, the data generated by the satellites are to be both developed and transmitted more quickly.

Since its inception, the GOES-R program has undergone several changes in cost and scope. As originally envisioned, GOES-R was to encompass four satellites hosting a variety of advanced technology instruments and providing 81 environmental products. The first two satellites in the series (called GOES-R and GOES-S) were expected to launch in September 2012 and April 2014.

However, in September 2006, NOAA decided to reduce the scope and technical complexity of the GOES-R program because of expectations that total costs, which were originally estimated to be $6.2 billion, could reach $11.4 billion. Specifically, NOAA reduced the minimum number of satellites from four to two, cancelled plans for developing an advanced instrument (which reduced the number of planned satellite products from 81 to 68), and divided another instrument into two separate acquisitions. The agency estimated that the revised program would cost $7 billion and kept the planned launch dates unchanged.

Subsequently, NOAA made several other important decisions about the cost and scope of the GOES-R program. In 2007, NOAA established a new program cost estimate of $7.67 billion and moved the launch dates for the first two satellites to December 2014 and April 2016. Further, to mitigate the risk that costs would rise, program officials decided to remove selected program requirements from the baseline program and treat them as contract options that could be exercised if funds allowed. These requirements included the number of products to be distributed, the time to deliver the remaining products (product latency), and how often these products would be updated with new satellite data (refresh rate). For example, program officials eliminated the requirement to develop and distribute 34 of the 68 envisioned products, including low cloud and fog, sulfur dioxide detection, and cloud liquid water. Program officials included the restoration of the requirements for the products, latency times, and refresh rates as options in the ground system contract that could be acquired at a later time. Program officials later reduced the number of products that could be restored as a contract option (called option 2) from 34 to 31 because they determined that two products were no longer feasible and two others could be combined into a single product.

In late 2009, NOAA changed the launch dates for the first two satellites to October 2015 and February 2017. More recently, NOAA restored two satellites to the program's baseline, making GOES-R a four-satellite program once again. In February 2011, as part of its fiscal year 2012 budget request, NOAA requested funding to begin development for two additional satellites in the GOES-R series—GOES-T and GOES-U. The program estimated that the development for all four satellites in the GOES-R series—GOES-R, GOES-S, GOES-T, and GOES-U—would cost $10.86 billion through 2036, an increase of $3.19 billion over its 2007 life cycle cost estimate of $7.67 billion for the two-satellite program.

Table 2. Key Changes to the Geostationary Operational Environmental Satellite-R Program over Time

	August 2006 (baseline program)	**September 2006**	**November 2007**	**February 2011**	**August 2013**
Number of satellites	4	2	2	4	4
Instruments	•Advanced Baseline Imager •Geostationary Lightning Mapper •Magnetometer •Space Environmental In-Situ Suite •Solar Imaging Suite (which included the Solar Ultraviolet Imager, and Extreme Ultraviolet/X-Ray Irradiance Sensor) •Hyperspectral Environmental Suite	•Advanced Baseline Imager •Geostationary Lightning Mapper •Magnetometer •Space Environmental In-Situ Suite •Solar Ultraviolet Imager •Extreme Ultraviolet/X-Ray Irradiance Sensor	No change	No change	No change
Number of satellite products	81	68	34 baseline 34 optional	34 baseline 31 optional	34 baseline 31 optional
Life cycle cost estimate (in then-year dollars)	\$6.2 billion – \$11.4 billion (through 2034)	\$7 billion (through 2028)	\$7.67 billion (through 2028)	\$10.86 billion (through 2036)[a]	\$10.86 billion (through 2036)
Estimated launch dates for GOES-R and GOES-S	GOES-R: September 2012 GOES-S: April 2014	GOES-R: September 2012 GOES-S: April 2014	GOES-R: December 2014 GOES-S: April 2016	GOES-R: October 2015 GOES-S: February 2017	GOES-R: by March 2016 GOES-S: by June 2017[b]

Source: GAO analysis of NOAA data. | GAO-15-60

[a]Based on NOAA's fiscal year 2012 budget estimate, \$7.64 billion of this cost estimate was for the first two satellites in the series, GOES-R and GOES-S. The cost for the remaining two satellites— GOES-T and GOES-U—was estimated at \$3.22 billion.

[b]Program documentation shows that the launch commitment dates were changed to the first quarter of 2016 and the second quarter of 2017, respectively. The launch dates in this chart reflect the latest month in which launch can occur and still meet the launch commitment dates.

In August 2013, the program announced that it would delay the launch of the first two satellites in the program, due in part to the effects of sequestration. Specifically, the launch of the GOES-R satellite was delayed from October 2015 to the quarter ending March 2016, and the expected GOES-S satellite launch date was moved from February 2017 to the quarter ending June 2017. See table 2 for an overview of key changes to the GOES-R program.

Program Structure

While NOAA is responsible for GOES-R program funding and overall mission success, it implemented an integrated program management structure with NASA for the GOES-R program since it relies on NASA's acquisition experience and technical expertise. The NOAA-NASA Program Management Council is the oversight body for the GOES-R program, and is co-chaired by the NOAA Deputy Undersecretary for Operations and the NASA Associate Administrator. NOAA also located the program office at NASA's Goddard Space Flight Center.

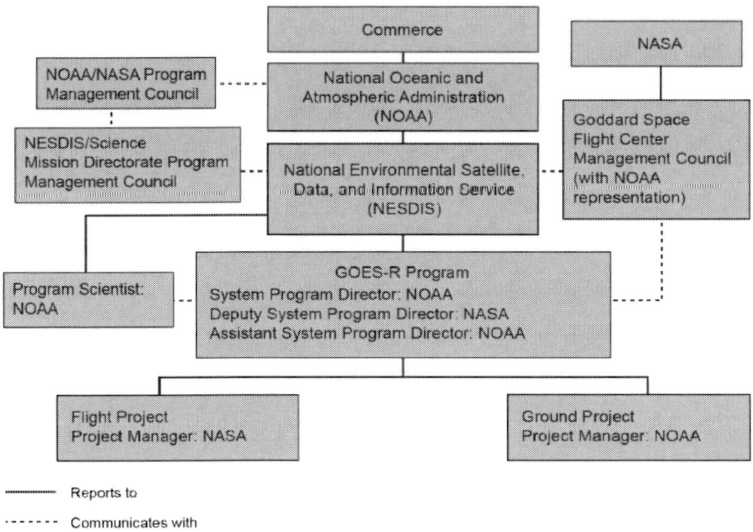

Source: NOAA. | GAO-15-60.

Figure 3. Organizational Structure and Staffing of the Geostationary Operational Environmental Satellite-R Series Program.

The GOES-R program is divided into flight and ground projects that have separate areas of responsibility and oversee different sets of contracts. The flight project, which is led by NASA, includes instruments, spacecraft, launch services, satellite integration, and on-orbit satellite initialization. The ground project, which is led by NOAA, is made up of three main components: the core ground system, an infrastructure of antennas, and a product access subsystem. In turn, the core ground system comprises four functional modules supporting operations, product generation, product distribution, and configuration control. Figure 3 depicts the integrated program management structure and the organization of the flight and ground projects within that structure, while table 3 summarizes the GOES-R instruments and their planned capabilities and table 4 describes key components of the ground project.

Table 3. Instruments on the Geostationary Operational Environmental Satellite-R Program

Instrument	Description
Advanced Baseline Imager	Expected to provide variable area imagery and radiometric information of the earth's surface, atmosphere, and cloud cover. Key features include • monitoring and tracking severe weather; • providing images of clouds to support forecasts; and • providing higher resolution, faster coverage, and broader coverage simultaneously.
Geostationary Lightning Mapper	Expected to continuously monitor total lightning (in-cloud and cloud-to-ground) activity over the United States and adjacent oceans and to provide a more complete dataset than previously possible. Key features include • detecting lightning activity as an indicator of severe storms and convective weather hazard impacts to aviation; and • providing a new capability to GOES for long-term mapping of total lightning that only previously existed on NASA low-earth-orbiting research satellites.
Magnetometer	Expected to provide information on the general level of geomagnetic activity, monitor current systems in space, and permit detection of magnetopause crossings, sudden storm commencements, and substorms.
Space Environmental In-Situ Suite	Expected to provide information on space weather to aid in the prediction of particle precipitation, which causes disturbance and disruption of radio communications and navigation systems. Key features include • measuring magnetic fields and charged particles;

Instrument	Description
	• providing improved heavy ion detection, adding low-energy electrons and protons; and • enabling early warnings for satellite and power grid operation, telecom services, astronauts, and airlines.
Solar Ultraviolet Imager	Expected to provide coverage of the entire dynamic range of solar X-ray features, from coronal holes to X-class flares, and will provide quantitative estimates of the physical conditions in the Sun's atmosphere. Key features include • providing information used for geomagnetic storm forecasts, and power grid performance; and • providing observations of solar energetic particle events related to flares.
Extreme Ultraviolet/X-Ray Irradiance Sensor	Expected to detect solar soft X-ray irradiance and solar extreme ultraviolet spectral irradiance. Key features include • monitoring solar flares that can disrupt communications and degrade navigational accuracy, affecting satellites, astronauts, high latitude airline passengers; and • monitoring solar variations that directly affect satellite drag/tracking and ionosphericchanges, which impact communications and navigation operations.

Source: GAO analysis of NOAA data. | GAO-15-60.

Table 4. Key Components of the Geostationary Operational Environmental Satellite-R Program Ground Project

Component	Description
Core Ground System	Expected to (1) provide command of operational functions of the spacecraft and instruments, (2) receive and process information from the instruments and spacecraft, (3) distribute satellite data products to users, and (4) provide configuration control and a common infrastructure and set of services for the satellite and instruments.
Antennas	Expected to provide six new antenna stations and modify four existing antennas to receiveGOES-R data. The antenna contract is also expected to include the construction of relatedinfrastructure, software development for control systems, and maintenance.
Product Distribution and Access System	Expected to provide ingestion of data and distribution for GOES-R products and data to authorized users. When completed, this system will be integrated into the core ground system.

Source: GAO analysis of NOAA data. | GAO-15-60.

Prior GAO Reports Made Recommendations to Address Program Weaknesses

In recent years, we issued a series of reports aimed at addressing weaknesses in the GOES-R program.[5] Key areas of focus included (1) cost, (2) technical challenges and changes in requirements, and (3) contingency plans.

Addressing cost risks: In June 2012, we reported that the GOES-R program might not be able to ensure that it had adequate resources to cover unexpected problems in remaining development.[6] We recommended the program strengthen its process for planning and reporting on reserves. More recently, in September 2013, we reported on weaknesses in the process for reporting reserves to management and recommended the agency take action to brief senior executives on a regular basis regarding the status of reserves.[7] The agency agreed with our recommendations and took steps to address them by identifying needed reserve levels and providing a more detailed breakdown of reserve percentage calculations. However, NOAA is not yet identifying the reserves associated with each satellite in the series.

- **Technical issues and changes in requirements:** We previously reported on issues related to GOES technical challenges and requirements. In 2012, we reported that key instruments were experiencing technical challenges and required additional redesign efforts. For example, emissions for the Geostationary Lightning Mapper instrument were outside the specified range. 8 Also, the ground project was experiencing ongoing technical problems—for example, the definition of ground system software requirements and integration of flight instruments. As a result, revisions were made to the Core Ground System's baseline development plan and schedule.
  More recently, in September 2013, we reported that NOAA made changes to several of the GOES-R requirements—including decreasing the accuracy requirement for the hurricane intensity product and decreasing the timeliness of the lightning detection product—and that end users were concerned by many of these changes.[9] We recommended that the program improve communications with users on changes in GOES-R requirements by assessing impacts, seeking input from users, and disseminating information on changes. NOAA agreed with this recommendation and

took steps to explore further avenues of user communication such as customer forums and interagency working groups.

- **Contingency planning:** In February 2013, due to the importance of environmental satellite data and the potential for a gap in this data, we added mitigating weather satellite gaps to our biennial High-Risk list.[10]

In that report, we noted that NOAA had established a contingency plan for a potential gap in the GOES program, but it needed to demonstrate its progress in coordinating with the user community to determine their most critical requirements, conducting training and simulations for contingency operations scenarios, evaluating the status of viable foreign satellites, and working with the user community to account for differences in product coverage under contingency operations scenarios. We also stated that NOAA should update its contingency plan to provide more details on its contingency scenarios, associated time frames, and any preventative actions it is taking to minimize the possibility of a gap.

More recently, in September 2013, we reported that, while NOAA had established contingency plans for the loss of the GOES satellites, these plans still did not address user concerns over potential reductions in capability, and did not identify alternative solutions and timelines for preventing a delay in the GOES-R launch date.[11] We recommended the agency revise the satellite and ground system contingency plans to address weaknesses, including providing more information on the potential impact of a satellite failure and coordinating with key external stakeholders on contingency strategies. The agency agreed with these recommendations and took steps to address them by identifying and refining program contingency plans.

NOAA is Making Progress on the GOES-R Satellite, but Schedule, Cost, and Functionality Challenges Remain

NASA and NOAA are following NASA's standard space system life cycle on the GOES-R program. This life cycle includes distinct phases, including concept and technology development; preliminary design and technology completion; final design and fabrication; system assembly, integration and testing, and launch; and operations and sustainment. Key program reviews are to occur throughout each of the phases, including preliminary design review, critical design review, and system integration review. NOAA and NASA jointly conduct key reviews on the flight and ground segments individually as well as for the program as a whole, and then make decisions on whether to proceed to the next phase. Figure 4 provides an overview of the life cycle phases, key program reviews, and associated decision milestones. In addition, the key reviews are described in table 5.

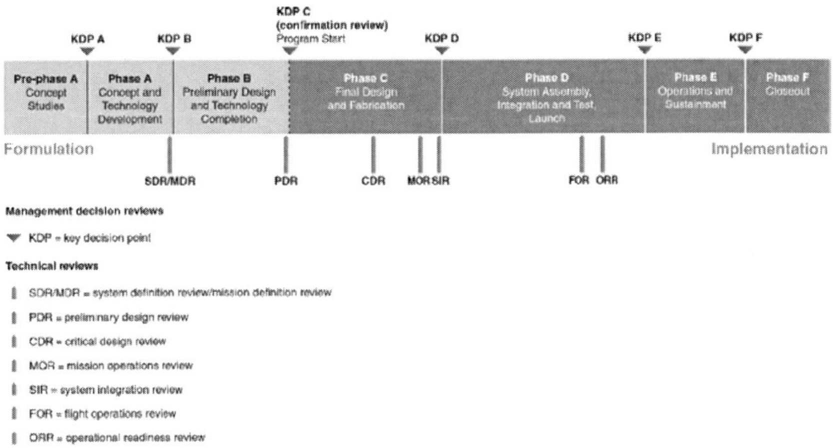

Source: GAO analysis of NASA data. | GAO-15-60.

Note: According to a NASA official, the MOR and FOR are considered lower-level reviews and are not mandated by NASA's primary procedural requirements. They are, however, key mission reviews required by NASA's Goddard Space Flight Center.

Figure 4. NASA's Life Cycle for Flight Systems.

The GOES-R program has completed important steps in developing its first satellite. Specifically, the program completed its critical design review in November 2012, its Mission Operations Review in June 2014, and its System Integration Review in July 2014. Based on the results of the System Integration Review, in September 2014, NOAA and NASA decided to move the program to the next phase, the system assembly, integration and test, and launch and checkout phase.

Table 5. Major Development Reviews for the Geostationary Operational Environmental Satellite-R Program

Review	Description
System Definition Review	Performed on the flight and ground segments individually, and then on the program as a whole, this review is to examine the proposed system architecture/design and demonstrate that a system that fulfills the mission objectives can be built within existing constraints.
Preliminary Design Review	Performed on the flight and ground segments individually, and then on the program as a whole, this review is to demonstrate that the preliminary design meets all system requirements with acceptable risk and within the cost and schedule constraints and to establish the basis for proceeding with detailed design.
Critical Design Review	Performed on the flight and ground segments individually, and then on the program as a whole, this review is to evaluate the completed detailed design of the element and subsystem products in sufficient detail to provide approval for a production stage.
Mission Operations Review	Performed programwide, this review is to establish the adequacy of plans and schedules for ground systems and flight operations preparation, and to justify readiness to proceed with implementation of the remaining required activities. It is typically held subsequent to completion of detail design and fabrication activity, but prior to initiation of major integration activities of flight or ground-system elements.
System Integration Review	Performed programwide, this review is to evaluate the readiness of the project to start system assembly, test, and launch operations. The objectives of the review include ensuring that planning is adequate for all remaining system activities and that available cost and schedule resources support completion of all necessary remaining activities with adequate margin.
Flight Operations Review	This review is to present the results of mission operations activities and show that the program has verified compliance with all requirements and demonstrated the ability to execute all phases and modes of mission operations, data processing, and analysis.
Operational Readiness Review	This review is to examine characteristics and procedures used in the system's operation and ensures that all system and support hardware, software, personnel, and procedures are ready foroperations and that user documentation accurately reflects the deployed state of the system. It is typically held near the completion of pre-launch testing between the flight segment and the ground system.

Source: GAO analysis of NOAA data. | GAO-15-60.

To prepare for the recent reviews and milestones, the program completed numerous important steps on both the flight and ground projects. Key accomplishments include:

- completing testing on individual components including the six flight instruments, the spacecraft core and system modules, and ground system components;
- releasing key ground system software components on enterprise infrastructure and mission management in December 2013 and April 2014, respectively, and completing the dry run of another ground system release that includes all planned products for the GOES-R satellite;
- replacing and successfully demonstrating a new engineering analysis tool, which will perform trending and offline analysis;
- completing installation and testing of antenna dishes at NOAA's satellite operations facility, continuing installation and testing at NOAA's primary satellite communications site, and beginning training in use of the antenna system;
- completing two key readiness reviews on the product distribution and access system; and
- completing connectivity tests throughout system hardware components.

Moving forward, the next major program milestones are the flight operations review and the operational readiness review. In preparation for these milestones, the program plans to conduct a series of five end-toend tests.[12] These tests are expected to validate compatibility between the space and ground segments before the launch of the first satellite.

Schedule Delays Continue and Could Affect the Planned Launch Date

While the program is making progress, it has experienced recent and continuing schedule delays. In 2013, we reported that technical issues on both the flight and ground projects had the potential to cause further delays to the program schedule, and that delays in interim key milestones could result in delays to the launch readiness date. 13 We also reported that the program had delayed many of these key milestones and tests, including the mission operations review and the five end-to-end tests.

In the months since that report, these and all other major milestones have been further delayed by 5 to 8 months. For example, the program delayed its mission operations review an additional 5 months to June 2014. This most recent delay occurred when the program found that detailed plans for future operations of the system were inadequate and behind schedule.

The GOES-R program cited multiple reasons for these recent delays. They include challenges in delivering software builds for the flight project and delays in completing communication testing for the spacecraft. In addition to these delays, as previously noted, in late 2013 the agency moved the launch commitment date of the first GOES satellite from October 2015 to the first quarter of 2016. In September 2014, a NOAA management council also confirmed the agency's commitment to launch the satellite during that quarter. Table 6 and figure 5 show changes in the planned or completed dates of selected key program milestones over time.

The program has experienced recent issues in developing specific components that increase the potential for further delays. For example, in mid-2014, the program discovered an issue in newly-designed electronics systems to be used on the spacecraft system module. According to the program, this issue could cause a schedule delay because it was discovered late in the integration and testing phase. As we have previously reported, experiencing problems during the integration and test phase often lead to cost and schedule growth.[14] Also, one of the main ground antenna sites experienced power supply failures on 10 separate occasions, due to issues in the site's air conditioning units. According to the program officials, this issue delayed the completion of testing.

In addition, the program's actions to mitigate schedule delays introduce some risks, and could therefore increase the amount of the delay. For example, the program attempted to mitigate delays in developing the detailed plans for ground-based data operations by performing system development while concurrently working on the detailed plans. According to program documentation, this could cause rework and increase cost and schedule slips. In addition, the program has responded to prior delays by eliminating selected repetitive tests and moving to a 24-hour-aday, 7-day-a-week spacecraft integration testing schedule. These actions increase the risk that it will take longer to identify issues and that there will be no extra time available to expand testing into off hours, if needed.

Table 6. Delays in Milestones for the Geostationary Operational Environmental Satellite-R Program

Program milestone	Date planned (as of April 2012)	Date completed or planned (as of March 2013)	Delay through March 2013	Date completed or planned (as of November 2014)	Additional delay (from April 2013 to November 2014)	Total delay from April 2012 to November 2014
Mission operations review	January 2013	January 2014	12 months[a]	June 2014	5 months	17 months
End-to-end test #1	February 2014	May 2014	3 months	December 2014	7 months	10 months
End-to-end test #2	May 2014	August 2014	3 months	March 2015	7 months	10 months
End-to-end test #3	August 2014	December 2014	4 months	May 2015	5 months	9 months
Flight operations review	September 2014	January 2015	4 months	July 2015	6 months	10 months
End-to-end test #4	December 2014	March 2015	3 months	September 2015	6 months	9 months
Operational readiness review	April 2015	April 2015	No change	December 2015	8 months	8 months
End-to-end test #5	July 2015	July 2015	No change	January 2016	6 months	6 months
Anticipated launch	October 2015	October 2015	No change	March 2016	5 months	5 months

Source: GAO analysis of NOAA data. | GAO-15-60

[a]Program officials stated that they had erroneously scheduled the mission operations review too soon, and moved the date by 9 months to better reflect when the review was needed. Therefore, only 3 of the 12 months were attributable to a delay.

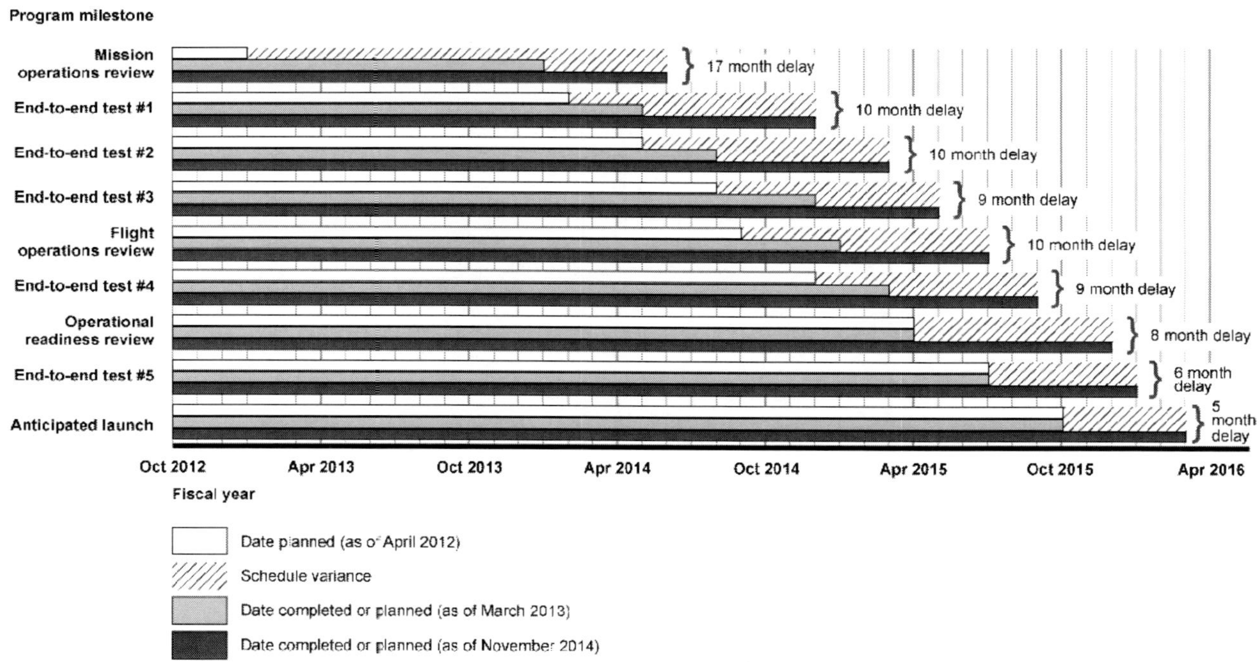

Source: GAO analysis of NOAA data. | GAO-15-60.

Figure 5. Changes in Key Test Dates for the Geostationary Operational Environmental Satellite-R Program over Time.

We have previously reported that overlapping planning and development activities and compressing test schedules are activities that increase the risk of further delays because there would be little time to resolve any issues that arise.[15] Program officials acknowledged that there was the possibility of further schedule slips.

The program is now reaching a point where additional delays in starting the end-to-end tests could begin to adversely affect its schedule. Recently, a NOAA review board reported that moving all of the end-toend and data operations tests further back into 2015 would cause a situation where too much testing would need to occur at once. In addition, any further delays could affect the committed launch date of the first GOES satellite. As of August 2014, program officials could not rule out the possibility of further delays in the planned launch date.

Program Life Cycle Cost Estimate Remains Relatively Steady, but Key Components are Costing More than Expected

NOAA's estimated life cycle cost for the GOES-R program has held relatively steady. NOAA's current life cycle cost estimate for the GOES-R program is $10.83 billion, which is slightly less than the $10.86 billion lifecycle cost estimate from August 2013. The $30 million change in the life cycle cost estimate from last year is the net result of moving selected management functions outside the program and addressing program commitments impacted by funding reductions associated with sequestration in late 2013. However, program data show that individual components are costing more than expected.[16]

Earned Value Management Data Show Key Components are Costing More than Expected

Federal agencies and private industry organizations often implement earned value management (EVM) as a tool for ensuring work completed is on track with expected costs and schedules.[17] EVM is a project management tool that, among other things, produces early warning signs of impending schedule slippages and cost overruns. Key EVM metrics include cost and schedule variances, which measure the value of work accomplished in a given period and compares it with the planned value of work scheduled for that period and with the actual cost of work accomplished. For example, an increase in cost variance means that the program spent more than expected to produce the work.

An analysis of EVM data for three key components—the GOES ground system, the ABI instrument, and the GLM instrument—shows that each experienced a growing cost variance. Specifically, over the twelve-month period ending July 2014, the cost variance for the ground system increased to 8.4 percent of total cumulative budgeted cost and for GLM, the cost variance increased to 5.1 percent. For a third key component, the ABI instrument, cost variance increased slightly from 2.4 to 2.6 percent. Figures 6, 7, and 8 show monthly cost variance data for the three key components for the year ending July 2014.

Program officials stated that cost variances during this period for the ground system were primarily due to increased labor costs needed to complete the contract deliverables on schedule. Because the program is moving into integration and testing—the phase of development in which cost growth is most likely[18]—cost variances will likely continue to increase. If these variances continue to increase, the program will need to use more of its contingency reserves to cover the costs, and the overall cost of the GOES-R program could increase.

Inconsistencies in EVM Data Make Oversight More Difficult

Best practices in cost estimation and monitoring call for agencies to confirm that data are reliable and valid, and that the calculations for each cost element are correct and the results make sense. [19] In reviewing EVM data for the GLM and ABI instruments we found inconsistencies in the contractor's monthly and cumulative reports that made it more difficult for NOAA to effectively oversee the contractor's performance. Specifically, we found inconsistencies between cumulative and monthly budget totals in contractor performance reports that ranged from hundreds of thousands to millions of dollars. For instance, the cumulative amount of budget allocated to complete work for GLM between February 2013 and March 2013 increased by just under $2 million, while the stated monthly change was $3.8 million. Also, the cumulative amount of budgeted work accomplished for ABI between May 2013 and June 2013 increased by $3.2 million, while the stated monthly change was $5.4 million. Month-tomonth discrepancies such as this occurred in each of 6 months for the GLM instrument and each of 9 months for the ABI instrument in the period between August 2013 and July 2014.

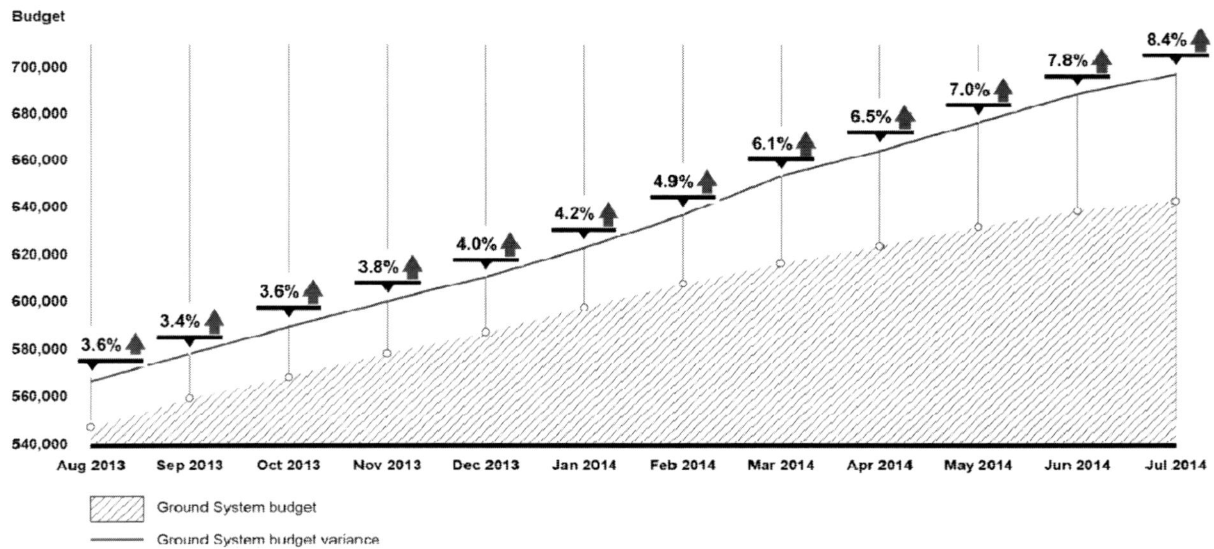

Source: GAO analysis of NOAA data. | GAO-15-60

Figure 6. Increases in Cost Variance for the Geostationary Operational Environmental Satellite-R Program's Ground System.

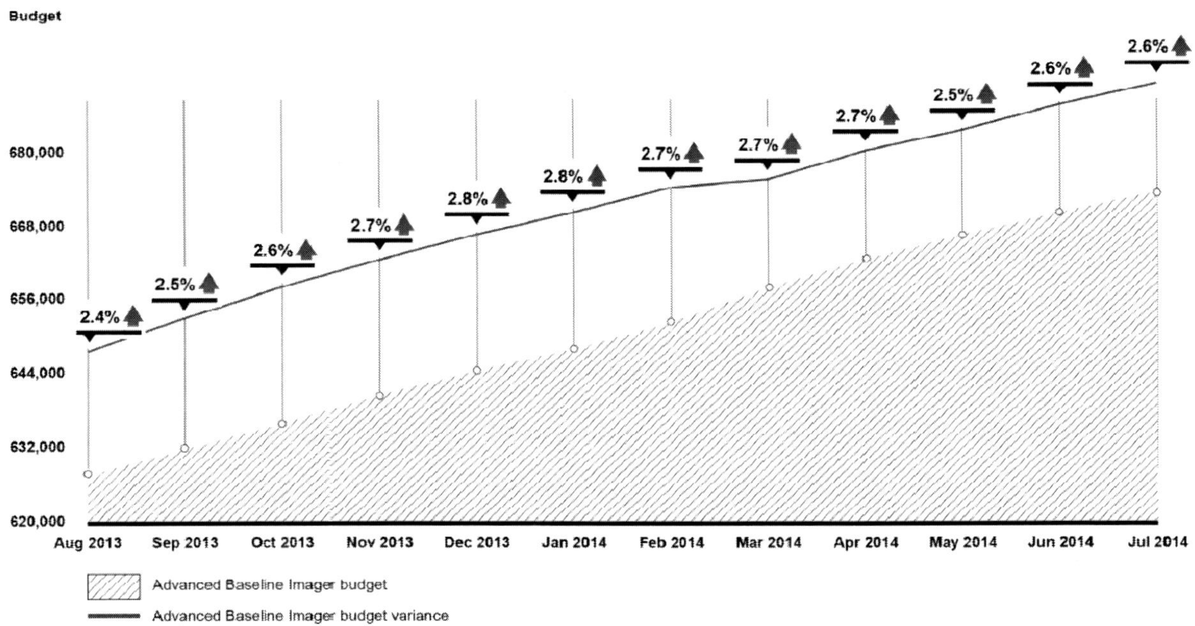

Source: GAO analysis of NOAA data. | GAO-15-60.

Figure 7. Increases in Cost Variance for the Geostationary Operational Environmental Satellite-R Program's Advanced Baseline Imager Instrument.

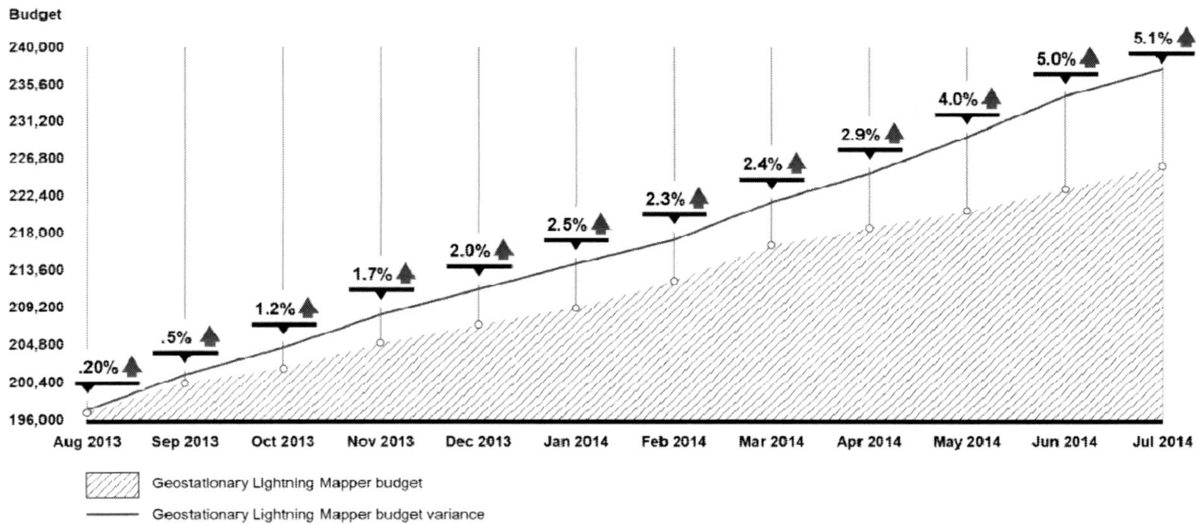

Source: GAO analysis of NOAA data. | GAO-15-60.

Figure 8. Increases in Cost Variance for the Geostationary Operational Environmental Satellite-R Program's Geostationary Lightning Mapper Instrument.

Program officials stated that these issues were addressed in later contractor reports, and that program analysts communicate with contractors and program management regularly to resolve any found discrepancies. However, more recent monthly reports continue to show discrepancies. If the instrument's cost data are unreliable, it is difficult for managers and program officials to make financial projections and assess reserve needs and usage.

The GOES-R Program is Considering Reducing or Deferring Functionality on the Ground System

The GOES-R program is considering eliminating or deferring planned functionality on its ground system due to issues experienced during development. Specifically, the GOES-R program is considering deferring functionality on the ground system in order to provide schedule relief in the case of further delays. Program and contracting officials recently revised the composition of the software releases it will be delivering on the ground system. In doing so, the GOES-R program identified "off-ramps," or decision points, at which time they could remove or defer a specific function from pre-launch to post-launch if it is not ready in time for testing.

As of September 2014, officials identified 50 potential decision points for deferring functionality. To date, the program has decided to implement five of the deferrals, including one to remove the ability to play back information from alternate ABI data sets outside the GOES ground system, and another to remove a low-level navigation capability. In addition, program officials decided against deferring 30 of the functions, leaving 16 deferrals that could be implemented in the future. The off-ramps still under consideration include a reduction in the amount of verification and validation activities that will be conducted.

NOAA Has Implemented a Defect Management Process, but Shortfalls Exist and Selected Critical Defects Remain Unresolved

A key element of a successful test phase is appropriately identifying and handling any defects or anomalies that are discovered during testing. Key aspects of a sound defect management process include defect management planning, defect identification and classification, defect analysis, defect resolution, defect tracking, and defect trending and reporting. Leading industry and government organizations consider defect management and resolution to be among the primary goals of testing.[20] These organizations have identified best practices for managing defects and anomalies that arise during system testing. Table 7 outlines the best practices of a sound defect management process.

Table 7. Best Practices in Managing Defects

Category	Best practice
Defect management planning	Establish procedures to maintain reasonable assurance that significant defects (and anomalies) will be discovered before production system deployment
	Ensure defect management is a part of standard testing procedures
	Establish procedures that link test exit criteria to resolution of defects
	Establish realistic schedules to discover and resolve defects during testing
	Establish defect metrics (number and severity of defects, number of defects closed, duration, etc.)
	Clearly define roles and responsibilities for testing staff and key stakeholders
Defect identification and classification	Classify each defect/anomaly using a series of descriptive attributes (type, severity, date of detection, etc.)
	Assign a priority or severity category to each defect/anomaly
Defect analysis	Assess impact of the defect on system operation and performance requirements, and that defects are traceable back to original requirements
	Analyze defects and determine their source and root cause
	Evaluate the impact of defects, and the steps required to mitigate them, on resources (time, cost)
Defect resolution	Decide how to address and resolve defects (make a fix, establish a workaround, require a waiver, etc.)
	Ensure that defect resolution complies with testing exit criteria
	Perform quality assurance to confirm the implementation of corrective actions, defect repairs, and preventive actions
Defect tracking	Track defect dispositions, rationale, and outcome of resolution activities
	Track defect-related metrics, such as number of defects open, closed, and duration (both in total and by severity)
	Monitor the risks of open high-priority defects throughout the project

Category	Best practice
Defect trending and reporting	Track and analyze defect trends over time, such as defect density and severity
	Report key defects to management and stakeholders
	Management should analyze information on reported problems, and should ensure the impact of any changes is determined, controlled, and monitored
	Report key defects to management and stakeholders

Source: GAO analysis of best practices identified by leading government and industry organizations. | GAO-15-60

The GOES-R program has sound defect management policies in place and it, along with its contractors, is actively performing defect management activities. Specifically, the program has fully satisfied 13 of the 20 best practices, and partially satisfied the remaining 7 practices. For example, the program has defect management procedures in place as part of its overall testing program. Defect management is incorporated in the program's mission assurance, configuration management, and verification/validation functions. In addition, for three key components we reviewed—ABI, GLM, and the ground system—defects are entered into automated systems, from which they are analyzed and resolved. The program also tracks, and regularly reports weekly and monthly, on defect totals and metrics to NOAA management.

However, there are several areas in which defect management policies and practices are inconsistent, including in performing and recording information pertinent to individual defects, and in reporting and tracking defect information. Table 8 provides an assessment of how the GOES-R program and key contractors performed on each of the best practices, and is followed by a more detailed discussion of shortfalls.

Table 8. Assessment of Geostationary Operational Environmental Satellite-R Program Practices in Managing Defects

Category	Best practice	Assessment	Description
Defect management	Establish procedures to maintain reasonable assurance that significant defects (and anomalies) will be discovered before production system deployment	●	NOAA has key procedures in place that provide assurance that defects will be discovered before system deployment. These include mission assurance, configuration management, and verification and validation procedures.

Table 8. (Continued)

Category	Best practice	Assessment	Description
			In addition, individual program components are required to establish procedures for reporting nonconformance, discrepancies, defects and anomalies.
	Ensure defect management is a part of standard testing procedures	●	Defect management is a part of NOAA's overall testing approach. For example, mission assurance and engineering management plans explain that testing results will be reviewed for both accuracyand compliance with mission objectives, and that processes and systems will be in place to report and resolve defects and anomalies. Further, NOAA established processes for reporting defects identified during testing.
	Establish procedures that link test exit criteria to resolution of defects	●	NOAA program procedures call for ensuring the traceability of requirements and mission objectives throughout the testing phase, and assign staff to perform oversight of testing. Program documentation traces exit criteria and results to mission level objectives to ensure changes are tracked and controlled, throughout the project lifecycle.

Category	Best practice	Assessment	Description
	Establish realistic schedules to discover and resolve defects during testing	◐	NOAA established a schedule for the GOES-R program's integration and testing phase, which includes time for defect discovery and resolution. However, the schedule was not realistic. The program has delayed major milestones several times due to issues discovered during testing and has compressed its overall testing period.
	Establish defect metrics (number and severity of defects, number of defects closed, duration, etc.)	◐	NOAA established metrics for tracking defects on its software components, including number and severity. However, NOAA did not establish requirements for metrics in tracking defects in its hardware components. Moreover, NOAA did not establish a consistent approach for classifying the priority and severity of defects. As a result, metrics and methodologies for classifying defects are not consistent between contractors or subcontractors.
	Clearly define roles and responsibilities for testing staff and key stakeholders	●	NOAA and contractor procedures establish and describe the key roles and responsibilities of individuals and groups that are involved in integration and testing phase activities.

Table 8. (Continued)

Category	Best practice	Assessment	Description
Defect identification and classification	Classify each defect/anomaly using a series of descriptive attributes (type, severity, date of detection, etc.)	●	All three components we reviewed performed this practice. For ABI, GLM, and the ground system, contractors fully classified each defect with a range of descriptive attributes such as a defect's status, description, and the activity during which it occurred.
	Assign a priority or severity category to each defect/anomaly	◐	For the ground system, contractors assigned a priority category for each defect or anomaly. For ABI and GLM, individual defect reports provide qualitative information on priority and severity, but there is no standard rating scale conversion for these attributes.
Defect analysis	Assess impact of the defect on system operation and performance requirements, and that defects are traceable back to original requirements	●	For all three components we reviewed, contractors discussed potential impact in individual defect reports and traced defects back to original requirements in cases where defects affected those requirements.
	Analyze defects and determine their source and root cause	●	For all three components, contractors analyzed defects and determined their root cause.
	Evaluate the impact of defects, and the steps required to mitigate them, on resources (time, cost)	◐	For the ground system, contractors evaluated the impact of defects and defect mitigation steps on resources.

Category	Best practice	Assessment	Description
			On the ABI and GLM instruments, contractors did not evaluate the impact of defects on resources. Moreover, NOAA did not provide evidence that it assessed the impact of key defects on resources.
Defect resolution	Decide how to address and resolve defects (make a fix, establish a workaround, require a waiver, etc.)	●	All three program components have a disposition or resolution plan in place in order to close each defect.
	Ensure that defect resolution complies with testing exit criteria	●	For all three program components, there are records of exit testing compliance for individual defects. Also, relevant stakeholders must sign off before a successful test result can be achieved.
	Perform quality assurance to confirm the implementation of corrective actions, defect repairs, and preventive actions	●	For all three program components, contractors provided evidence of quality assurance steps taken to address defects.
Defect tracking	Track defect dispositions, rationale, and outcome of resolution activities	●	All three components tracked dispositions, corrective actions, and outcomes of defects. This was done either in defect reports or at the level of defect review boards. Contractors also maintain additional analysis reports and tracking systems.

Table 8. (Continued)

Category	Best practice	Assessment	Description
	Track defect-related metrics, such as number of defects open, closed, and duration (both in total and by severity)	●	For the three components we reviewed, contractors tracked defect totals and metrics, such as number of defects open and closed. They also tracked information on the hardware or software severityrating for each defect.
	Monitor the risks of open high-priority defects throughout the project	●	For the three components we reviewed, contractors tracked and reviewed all open high-priority defects.
Defect trending and reporting	Track and analyze defect trends over time, such as defect density and severity	◐	NOAA's prime contractors on the flight and ground projects track, analyze, and provide trend data (including defect density) for both hardware and software defects. They also record information about defect root cause, lifecycle, and corrective actions. In addition, the contractors for flight instruments track defect metrics and statistics. However, because NOAA has not specified consistent metrics or defined how to assess the severity of defects, the contractor data does not lend itself to programwide analysis.
	Report key defects to management and stakeholders	◐	Both flight and ground project contractors report defect trend information and metrics to the GOES-R

Category	Best practice	Assessment	Description
			program and an independent NOAA mission assurance group monthly, and the mission assurance group reports on defects both weekly and monthly to NOAA management. However, the information reportedto NOAA is not comprehensive or comparable because each of the contractors maintains its own disparate reporting standards and metrics. For example, on the ground system, the contractor tracks and reports defect totals split out monthly by type and severity; on GLM, only software defects are reported monthly, with totals not split out by severity; and on ABI, high- and low-priority defects are tracked and reported in different ways and for different lengths of time.
	Management should analyze information on reported problems, and should ensure the impact of any changes is determined, controlled, and monitored	00	NOAA management provides oversight of key defects and the resolution of those defects during monthly program management meetings. However, managers were not briefed on all longstanding high-risk defects.

Table 8. (Continued)

Category	Best practice	Assessment	Description
			Specifically, none of the 14 high risk defects that that we assessed which had remained unresolved for more than 12 months were included in monthly briefings.

Source: GAO analysis of NOAA data. | GAO-15-60.

Among the shortfalls seen in table 8 are a number of cross-cutting themes:

- **Variation among contractors in managing and reporting metrics:** The GOES-R program affords its contractors and subcontractors wide latitude in making decisions on how to manage, track, and report defects, which results in variation among program components. For example, the program established a minimum set of metrics that must be reported and recorded for each software defect, but has not done so for hardware defects.
- **No clear definition of defects:** In its guidance, NOAA did not fully define the terminology of defect management. As a result, the program and contractors use terms such as defect, anomaly, nonconformance, incident reports, and trouble reports without explaining clearly how they are related to each other. Without understanding these relationships, variations from expected performance are likely to be treated differently throughout the program. For example, some issues that occur after formal integration and testing are complete are not considered as defects, which means that they are not all reported on the statistics provided to program management. In another case, an issue on instrument data algorithms was uncovered during user testing and required rework; however, it was not counted as a defect. Program officials stated that this was a documentation issue and noted that they do not track documentation issues as defects. However, the issue affected more than documentation. It was addressed by reworking the software algorithms; thus, it should have been tracked as a software defect.
- **No clear definition of priority/severity:** In its contract requirements and other guidance documents, the GOES-R program did not establish

specific guidance to its contractors on how to prioritize or establish the severity of defects. At most, documents list a severity classification as one of many defect attributes that contractors should review as necessary. A program official explained that, for hardware defects, there is no programwide policy for defect metrics, including priority or severity. As a result, the information tracked, trended, and reported to management on defect totals also varied by contractor, and thus often by instrument or component.

- **Unrealistic testing schedule:** Effective defect management requires a realistic schedule in that it takes time to be able to fully identify, analyze, prioritize, resolve, and track defects. However, in May 2014, an internal NOAA report stated that the current program testing schedule is "unrealistically aggressive." The GOES-R program has chosen to compress its remaining testing schedule, which increases the risk that there will not be sufficient time to address defects before moving to the next stage of testing. Effects of this type can already be seen; the program reported that some defects that were open for a long period of time have been delayed due to the need for all available resources to be applied in resolving new, current defects.
- **Limited trend analysis:** While the program and NOAA's mission assurance team have analyzed contractor-provided trends in the volume and severity of defects over time for the ground system and spacecraft, they do not routinely analyze trend data for all components. For example, on the ABI and GLM instruments, the program and mission assurance team do not analyze trend data monthly for hardware defects. Instead, these groups only assess individual defect reports for this subset of defects. It is important for the program to assess defect trends over time in order to provide to management a more complete picture of testing status. Assessing trends in defect handling can result in better resource allocation to components of greatest need, and more attention to the effectiveness of resolving defects.

Although the GOES program has defect management policies in place and its contractors are actively tracking, analyzing, and reporting on defects, the discrepancies between contractors' data could cause issues in completing the remainder of the program's integration and testing period. For example, without consistent defect metrics, it is more difficult for managers to obtain a complete picture of the status of open, closed, and high priority defects across the program. Program officials stated that the program did not contractually specify any best practices, but that it assumed that contractors with high-level professional certifications would employ all practices necessary for the development effort. Unless the program can find a way to unite these disparate approaches to provide consistency in the methods by which defects are identified, prioritized, captured, and tracked, it will be more difficult for management to analyze and understand trends in opening and closing high-priority defects or to make decisions on how to best resolve the defects. Moreover, until the program addresses shortfalls in its defect management processes, it may not have a complete picture of remaining defects and runs the risk of not having sufficient time to resolve them.

Data Show a High Number of Unresolved Defects for Selected Key Components

While effectively and efficiently addressing and resolving defects is an important part of a sound system testing approach, the GOES-R program has not efficiently closed defects on selected components. As noted earlier, the program does not obtain or maintain defect trend data for the program as a whole; however, data on individual components and portions of components show that a large number of defects remain open, including several high-priority defects. Specifically, data for the GOES ground system shows that 500 defects remained open as of September 2014, including 36 high-priority defects. Defect data for the spacecraft show that it is taking an increasing amount of time to close hardware-related defects, but that the program is making progress in closing software defects. Specifically, as of April 2014, 42 software and 332 hardware defects were unresolved.

Defect totals on GOES instruments declined as the program approached the deadline for them to be completed in order to be integrated with the spacecraft. Specifically, for the GLM instrument, the total number of both hardware and software defects declined. Hardware defects declined from 117 in May 2014 to 13 in September 2014, and there were no remaining software defects by January 2014. For the ABI instrument, the number of newly opened hardware and software defects each month has declined over time, with only one unresolved defect remaining in July 2014. Table 9 depicts summary information on defects for selected components at different points in time. In addition, appendix II provides more information on defect trends for these components.

Table 9. Summary of Defects for Selected Geostationary Operational Environmental Satellite-R Program Components

Component	Total defects	Resolved defects	Unresolved defects	Unresolved high-priority defects
Ground Segment, as of September 2014	5686	5186	500	36
Spacecraft (software), as of April 2014	3420	3378	42	Not identified
Spacecraft (hardware), as of April 2014	Not identified	Not identified	332	Not identified
Geostationary Lightning Mapper (software), as of January 2014	54	54	0	0
Geostationary Lightning Mapper (hardware), as of September 2014	152	139	13	Not identified
Advanced Baseline Imager, as of July 2014	333	332	1	Not identified

Source: GAO analysis of NOAA data. | GAO-15-60

Program officials noted that in some cases, due to time concerns, lower-priority defects which have been determined to not have a major effect on performance are not closed until later in the testing process. Also, program officials have stated that they are having difficulty in closing defect-related incident reports due to insufficient manpower. Until the program reduces the number of longstanding unresolved defects, it faces an increased risk of further delays to the GOES-R launch date should an open defect affect future performance.

Facing a Gap in Backup Satellite Coverage, the GOES-R Program Has Improved Contingency Plans Though Shortfalls Remain

GOES satellite data are considered a mission-essential function because of their criticality to weather observations and forecasts. These forecasts—such as those for severe storms, hurricanes, and tornadoes— can have a substantial impact on our nation's people, infrastructure, and economy. Because of the importance of GOES satellite data, NOAA's policy is to have two operational satellites and one backup satellite in orbit at all times. This policy proved useful in December 2008 and again in September 2012, when the agency experienced problems with one of its operational satellites, but was able to move its backup satellite into place until the problems had been resolved.

However, NOAA is facing a period of up to 17 months when it will not have a backup satellite in orbit. Specifically, in April 2015, NOAA expects to retire one of its operational satellites (GOES-13) and to move its backup satellite (GOES-14) into operation. Thus, the agency will have only two operational satellites in orbit—and no backup satellite—until GOES-R is launched and completes an estimated 6-month post-launch test period. If GOES-R is launched in March 2016, the earliest it could be available for operational use would be September 2016. Figure 9 shows the potential gap in backup coverage, based on the launch and decommission dates of GOES satellites.

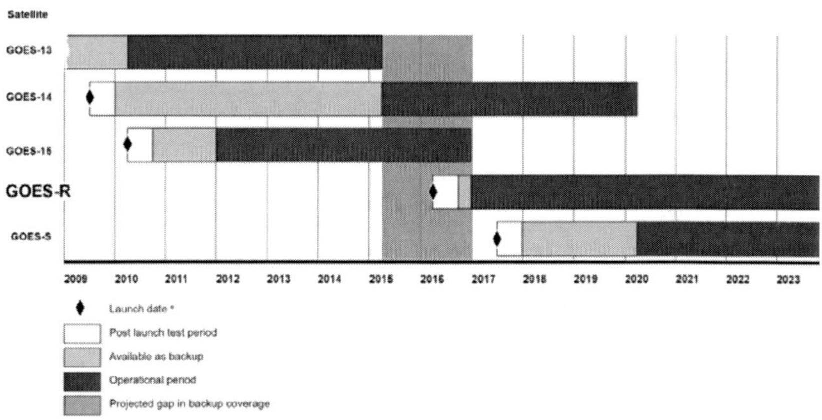

Source: GAO analysis of NOAA data. | GAO-15-60.

[a]The GOES-R and GOES-S launch dates reflect the end of the quarters listed in NOAA's latest launch estimates.

Figure 9. Potential Gap in Geostationary Operational Environmental Satellite Coverage, as of April 2014.

During the time in which no backup satellite would be available, there is a greater risk that NOAA would need to either rely on older satellites that are beyond their expected operational lives and may not be fully functional, request foreign satellite coverage, or to operate with only a single operational satellite. Agency officials stated that the risk of a gap may be reduced, because NOAA satellites have historically remained in operation longer than their expected life. While many satellites outlive their expected lifespans, the current GOES satellites are operating with reduced functionality, and one has experienced two major outages. Without a full complement of operational GOES satellites, the nation's ability to maintain the continuity of data required for effective weather forecasting could be compromised. This, in turn, could put the public, property, and the economy at risk.

Any delay to the GOES-R launch date would extend the time without a backup to more than 17 months. As discussed earlier in this report, further delays to the committed launch date of the first GOES-R satellite are possible due to continued technical issues encountered during testing and integration.

GOES-R Satellite Contingency Plan Shows Improvement, but Continues to Lack Details in Key Areas

Government and industry best practices call for the development of contingency plans to maintain an organization's essential functions—such as GOES satellite data—in the case of an adverse event.[21] In September 2013, we reported on weaknesses in the contingency plans for NOAA's satellites.[22] At that time, we compared NOAA's plans to 17 best practices associated with three main areas: identifying failure scenarios and impacts, developing contingency plans, and validating and implementing contingency plans. We reported that while NOAA identified failure scenarios, recovery priorities, and minimum levels of acceptable performance, the satellite contingency plan contained areas that fell short of best practices, such as working with the user community to account for potential reductions in capability under contingency operations. Furthermore, the agency did not identify alternative solutions or timelines for preventing a GOES-R launch delay.

In February 2014, NOAA released a new satellite contingency plan in response to these recommendations. This plan improved in comparison to many, but not all, of the best practices. Specifically, the plan improved in 6 areas and stayed the same in 4 areas. Table 10 compares our assessment of the current satellite contingency plan with our September 2013 analysis for all best practices that were not fully met.

Table 10. Assessment of Satellite Contingency Plans for the Geostationary Operational Environmental Satellite-R Program over Time

Category	Key element	September 2013 assessment	September 2014 assessment	Description
Identifying failure scenarios and impacts	Define likely failure scenarios	Fully implemented	Not applicable	
	Conduct impact analyses showing impact of failure scenarios on business processes and user requirements	Not implemented	Partially implemented	The plan now requires the NOAA Satellite Operations Command Center to notify users regarding impacts and outages. However, it does notspecify if or how impact analyses should be conducted.
	Define minimum acceptable level of outputs and recovery time objectives, and establish resumption priorities	Fully implemented	Not applicable	
Developing contingency plans	Define roles and responsibilities for implementing contingency plans	Partially implemented	Fully implemented	For each contingency scenario in the plan, roles and responsibilities are prominently listed. Detailed information on responsibilities are provided for each action in the plan.
	Identify alternative solutions to address failure scenarios	Partially implemented	Partially implemented	Solutions for each failure scenario are listed, including activation of an on-orbit spare or use of a foreign satellite, and shortening

Category	Key element	September 2013 assessment	September 2014 assessment	Description
				the on-orbit test period of a recently launched satellite. However, the plan does not address alternative solutions for preventing a launch delay.
	Select contingency strategies from among alternatives based on costs, benefits, and impacts	Partially implemented	Partially implemented	The satellite plan describes specific alternative actions that would be implemented depending on thecircumstances of a scenario, including the conducting of impact analyses and reporting procedures. However, the plan does not address strategies for preventing a delay in launch.
	Develop "zero-day" procedures	Partially implemented	Fully implemented	Zero-day events are defined, as are procedures for action if a zero-day event occurs. A subordinate policy document called out in the satellite contingency plan describes in more detail the actionsthat will be taken immediately should an event occur,and states that "at no point should immediate fix actions… be delayed."

Table 10. (Continued)

Category	Key element	September 2013 assessment	September 2014 assessment	Description
	Define actions needed to implement contingency strategies	Partially implemented	Fully implemented	The satellite plan describes detailed steps and actions for each major contingency strategy, including which groups should perform each action.
	Define and document triggers and timelines for enacting the actions needed to implement contingency plans	Partially implemented	Partially implemented	The satellite plan is specific in giving timelines and triggers for actions. For example, the plan calls for user notification procedures and a transition to full-disk imagery to begin after one hour of data loss, andother steps within 24 and 48 hours. However, the plan does not discuss triggers or timelines related to preventing a delay in satellite launch.
	Ensure that steps reflect priorities for resumption of products and recovery objectives	Partially implemented	Fully implemented	The satellite plan describes NOAA priorities and that of its component office, the National Environmental Satellite Data and Information Service.

Category	Key element	September 2013 assessment	September 2014 assessment	Description
				The plan then describes how the actions in the remainder of the plan are related to those priorities, in particular "restor(ing) the health of the satellite, and/or the flow of data." The plan also shows objectives for recovery of service.
	Designated officials review and approve contingency plan	Fully implemented	Not applicable	
Validating and implementing contingency plans	Identify steps for testing contingency plans and conducting training exercises	Fully implemented	Not applicable	
	Prepare for and execute tests	Fully implemented	Not applicable	
	Execute applicable actions for implementation of contingency strategies	Fully implemented	Not applicable	
	Validate test results for consistency against minimum performance levels	Partially implemented	Partially implemented	The satellite plan describes a system of regular test maneuvers performed each year. The plan also describes how these actions would be used in the event of an anomaly in order to meet operational priorities.

Table 10. (Continued)

Category	Key element	September 2013 assessment	September 2014 assessment	Description
				However, there is no indication that any of the metrics included in the plan are considered minimum performance levels in the event of an anomaly.
	Communicate and coordinate with stakeholders to ensure that contingency strategies remain optimal for reducing potential impacts	Partially implemented	Fully implemented	The satellite plan details methods for communicatingand coordinating with stakeholders, including actionsand responsibilities for notifying users in the event of an anomaly that would affect performance. A subordinate policy document called out in the satellite plan provides greater detail on expected user communications, stating that the Office of Satellite Product Operations should facilitate a complete and accurate data exchange among all parties on the potential effects of an anomaly.

Category	Key element	September 2013 assessment	September 2014 assessment	Description
				This plan also stated that typically within 2 hours, the office should inform stakeholders of unplanned interruptions to mission critical ground system IT services, product outages or product delays or reductions in service quality.
	Update and maintain contingency plans as warranted	Fully implemented	Not applicable	

Source: GAO analysis of NOAA documentation. | GAO-15-60

Program officials stated that it is not feasible to include strategies to prevent delays in launch of the first GOES-R satellite in the contingency plan, because such strategies are not static. They explained that options for preventing a delay vary greatly over time based on issues that occur during the satellite's development. They further stated that the program's monthly presentations to NOAA management provide a summary of current threats to the launch date and strategies to mitigate those threats. For example, NOAA is considering whether or not to remove or delay selected ground system functions to post-launch, and the program provides monthly updates on this issue.

While actively managing the program to avoid a delay is critical, it is also important that NOAA management and the GOES-R program consider and document feasible alternatives for avoiding or limiting such a launch delay. This will allow stakeholders throughout NOAA to be aware of, respond to, and plan for the potential implementation of each alternative, not only the small number of alternatives the program is actively considering in any given month. Until NOAA addresses the remaining shortfalls in its GOES-R gap mitigation plan, the agency cannot be assured that it is exploring all alternatives or that are able to effectively prepare to receive GOES information in the event of a failure.

CONCLUSION

After spending 10 years and just over $5 billion, the GOES-R program is nearing the launch of its first satellite. However, it continues to face challenges in maintaining its schedule and controlling its costs. The program continues to experience delays in remaining major milestones, which could result in further delays to the launch date. Costs are increasing faster than expected for key program components[23] and contractor data is often inconsistent from month to month. Until the agency ensures its contractor cost data are consistent, it will be more difficult for managers and program officials to make financial projections and assess reserve needs and usage.

As the GOES-R program progresses through its testing and integration phase, it is essential that it is able to appropriately handle defects that arise during testing. While NOAA and its contractors have implemented a defect management process which is successful in many areas, there are shortfalls in how the program defines defects, monitors trends, and reports on defects and defect metrics. In particular, NOAA has not established a standard set of metric information for all defects, including the dates on which defects were identified and resolved, or an indication of a defect's severity. Also, NOAA did not clearly define the type of issue that constitutes a defect. Furthermore, multiple defects remain open on some program components, including several that have remained open for more than six months. Until the program addresses these shortfalls and reduces the number of open defects, it may not have a complete picture of remaining issues and faces an increased risk of further delays to the GOES-R launch date.

NOAA could experience a gap in satellite data coverage if GOES-R is delayed further and one of the two remaining operational satellites experiences a problem. NOAA has made improvements to its satellite contingency plan, but the plan still does not sufficiently address mitigation options for a launch delay, potential impacts, or minimum performance levels. Until such information is available, it will be difficult to integrate mitigation efforts, or to coordinate with users in the event of a failure.

Recommendations for Executive Action

To address risks in the GOES-R program development and to help ensure that the satellite is launched on time, we are making the following four recommendations to the Secretary of Commerce. Specifically, we recommend that the Secretary of Commerce direct the NOAA Administrator to:

- investigate and address inconsistencies totaling hundreds of thousands of dollars in monthly earned value data reporting for the GLM and ABI instruments;
- address shortfalls in defect management identified in this report, including the lack of clear guidance on defect definitions, what defect metrics should be collected and reported, and how to establish a defect's priority or severity; and
- reduce the number of unresolved defects on the GOES ground system and spacecraft.

In addition, because NOAA has not fully implemented our prior recommendation to improve its satellite gap mitigation plan, we recommend that the Secretary of Commerce direct the NOAA Administrator to:

- add information to the GOES satellite contingency plan on steps planned or underway to mitigate potential launch delays, the potential impact of failure scenarios in the plan, and the minimum performance levels expected under such scenarios.

AGENCY COMMENTS AND OUR EVALUATION

We sought comments on a draft of our report from the Department of Commerce and NASA. We received written comments from the Deputy Secretary of Commerce transmitting NOAA's comments. NOAA concurred with all four of our recommendations and identified steps that it plans to take to implement them. It also provided technical comments, which we have incorporated into our report, as appropriate.

On November 14, 2014, an audit liaison for NASA provided an e-mail stating that the agency would provide any input it might have to NOAA for inclusion in NOAA's comments.

David A. Powner
Director, Information Technology Management Issues

APPENDIX I: OBJECTIVES, SCOPE, AND METHODOLOGY

Our objectives were to (1) assess progress on the GOES-R program with respect to planned schedule, cost, and functionality; (2) assess efforts to identify and address issues discovered during integration and testing; and (3) evaluate the likelihood of a gap in satellite coverage and analyze the adequacy of contingency actions in place to prevent or mitigate such a gap.

To assess NOAA's progress on the GOES-R program with respect to planned schedule, cost, and functionality, and to identify risks that could lead to further schedule delays, we analyzed data from monthly program management meetings. We evaluated progress made in completing key program components and major program reviews, and compared planned and actual completion dates for key program milestones over the last two to three years to determine the degree to which these dates have changed. To ensure that the program's schedule data were consistent and reliable, we compared milestone data over several months and contacted agency officials to corroborate events that occurred over the course of our engagement. We analyzed earned value management (EVM) data to compare levels of cost variance over time for key program components, and calculated earned value management metrics using program cost performance reports. To ensure that the program's cost data were reliable, we compared formulas from the GAO Cost Guide[1] to the program's EVM approach. We also compared EVM data across a series of monthly program cost performance reports. In doing so, we found inconsistencies in the monthly EVM data for selected components, and reported on those inconsistencies in this report. However, the data were sufficient for our purpose of assessing overruns because these inconsistencies were small in comparison to cost variances. We assessed recently enacted and potential changes in functionality for key program components. We also interviewed program officials regarding changes in schedule milestones, cost performance and reserve funding, and functionality.

To assess efforts to identify and address issues discovered during integration and testing, we identified best practices in defect management from leading industry and government organizations.[2] We compared NOAA policy documents and defect management artifacts by the GOES program, its contractors, and an independent NOAA mission assurance group to the best practices. We selected three components, which are critical to the program's mission requirements, and evaluated recent test results and defects.[3] Specifically, we identified two recent tests for each component and analyzed artifacts associated with defects identified during those tests. We compared the defects against each best practice to determine if NOAA and its contractors fully implemented, partially implemented, or did not implement the best practices. The agency and contractor had to meet all aspects of the best practice to achieve a fully implemented score, some aspects to achieve a partially implemented score, and no aspects to achieve a not implemented score. To identify the program's recent performance in closing and managing defects, we analyzed basic defect trend information—such as number of defects opened and closed, and defect severity—for each of several program components. We developed charts to demonstrate defect trends. To ensure the data were reliable, we compared defect data from individual defect reports to agency trend charts, and sought corroboration from agency and contractor officials. We also interviewed agency and contractor officials to discuss their defect management processes and practices, and to confirm information we found while analyzing individual defect reports and trend charts.

To evaluate the likelihood of a gap in satellite coverage, we reviewed monthly program management presentations and other review board documentation. To analyze the adequacy of contingency actions in place to prevent or mitigate such a gap, we compared NOAA's latest satellite contingency plan to best practice criteria from industry and government.[4] We focused specifically on the ten areas we identified as weak in our prior report.[5] For each of the ten areas, we rated NOAA's contingency plan as having partially or fully implemented the best practice criteria. In addition, we interviewed agency officials regarding the likely amount of a gap in on-orbit backup coverage of GOES satellites, the potential for a gap in operational coverage, and efforts to improve the GOES contingency plan.

We conducted this performance audit from January 2014 to December 2014, in accordance with generally accepted government auditing standards. Those standards require that we plan and perform the audit to obtain sufficient, appropriate evidence to provide a reasonable basis for our findings and conclusions based on our audit objectives. We believe that the evidence obtained provides a reasonable basis for our findings and conclusions based on our audit objectives.

APPENDIX II: DEFECT TRENDS FOR KEY GOES-R COMPONENTS

As noted earlier, the GOES-R program does not obtain or maintain defect trend data for the program as a whole; however, data on individual components and portions of components show that a large number of defects remain open, including several high-priority defects. The following sections provide more details on recent data trends for the GOES-R ground system and spacecraft as well as the ABI and GLM instruments. It is important to note that the sections do not provide consistent information among components because the program does not require consistent metrics and the contractors document different data.

Ground System Has Many Unresolved Defects

As of the end of September 2014, the ground system had 500 open defects, an increase from 342 open defects at the end of September 2013. Also, as of the end of September 2014, 96 defects that were identified prior to

January 2014 remained open, and 36 high-priority defects (those rated as critical or moderate) remained open.

Program and contractor officials provided rationale and insight into the mitigating circumstances surrounding these defects. Program officials stated that the number of open defects during the integration and test phase is as expected for a ground system of a magnitude similar to GOES. Contractor officials reported that none of the 96 longstanding defects were in either of the two highest severity categories. They also noted that most of the open high-priority defects had been opened during testing events conducted over the previous 3 months, which is a reasonable length of time to close a defect.

Furthermore, according to contractor officials, many of the longstanding open defects were in categories such as documentation-related defects which are considered less severe. Officials also stated that defects in these categories are often kept open for specific reasons, such as to gain cost efficiencies or to wait for an already-planned test event rather than creating a new test event. However, as of September 2013, 193 of the 342 defects were not related to documentation and this number rose to 416 of the 500 open defects in September 2014. Figure 10 shows the increase in open defects in the period from September 2013 to August 2014 for the GOES-R ground system.

Spacecraft Trends Show Increasing Time to Address Hardware Defects While the Number of Software Defects Declined

As allowed by NOAA's guidance, contractors on the spacecraft track defects differently than the ground system contractors do. Thus, it is not possible to compare opened and closed defects over time as depicted above for the ground system. Instead, spacecraft contractors track hardware and software defects independently. They also track when new defects are identified and how long they stay open.

For spacecraft hardware, the defect data show that the average time it took to resolve defects increased over time. Specifically, the average age for hardware defects increased from 99 days in May 2013 to a high of 167 days in April 2014, at which point 58 percent of all open defects had been open for at least 90 days. Program officials stated that they began continuous 24-hour a day, 7-day a week testing in October 2014, which means that they should be able to make progress in closing the remaining defects.

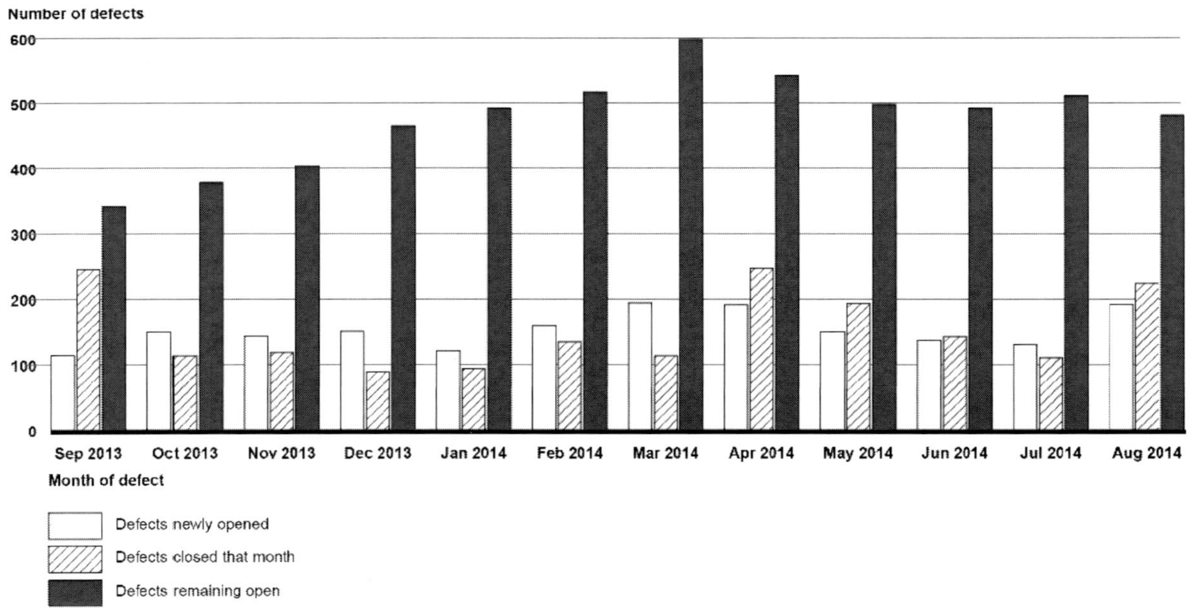

Source: GAO analysis of NOAA data. | GAO-15-60.

Figure 10. Number of Open and Closed Defects for the Geostationary Operational Environmental Satellite-R Program's Ground System, as of August 2014.

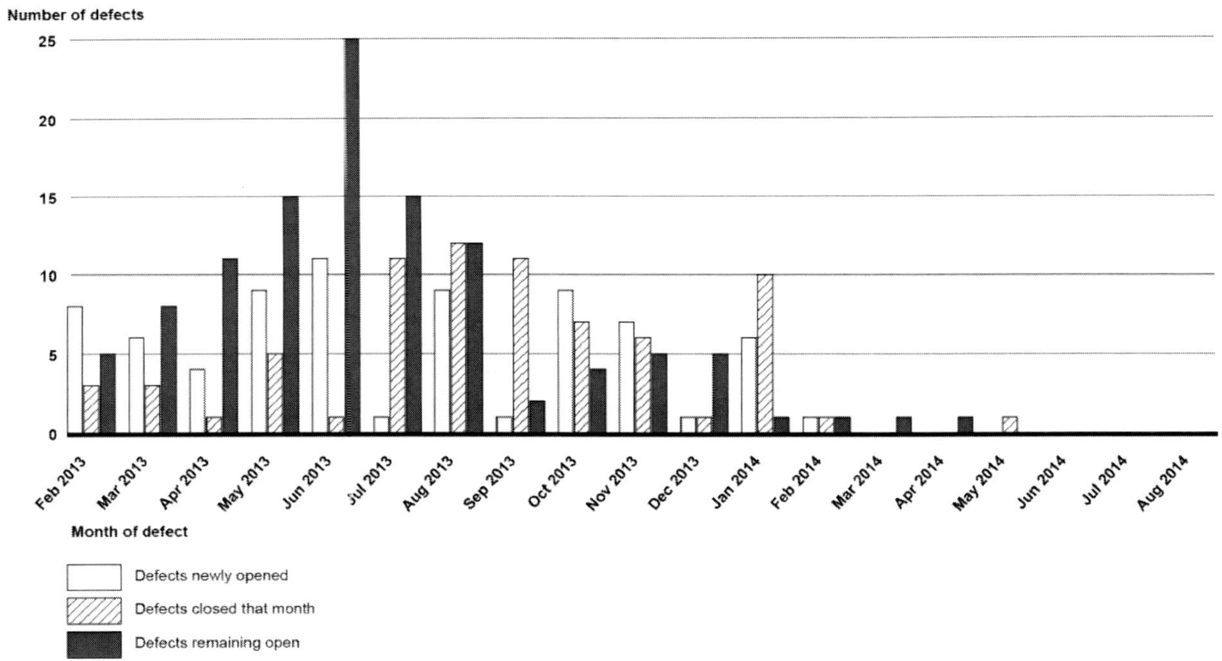

Source: GAO analysis of NOAA data. | GAO-15-60.

Figure 11. Number of Open and Closed Hardware Defects for the Geostationary Operational Environmental Satellite-R Program's Advanced Baseline Imager Instrument for Each Month from February 2013 to August 2014.

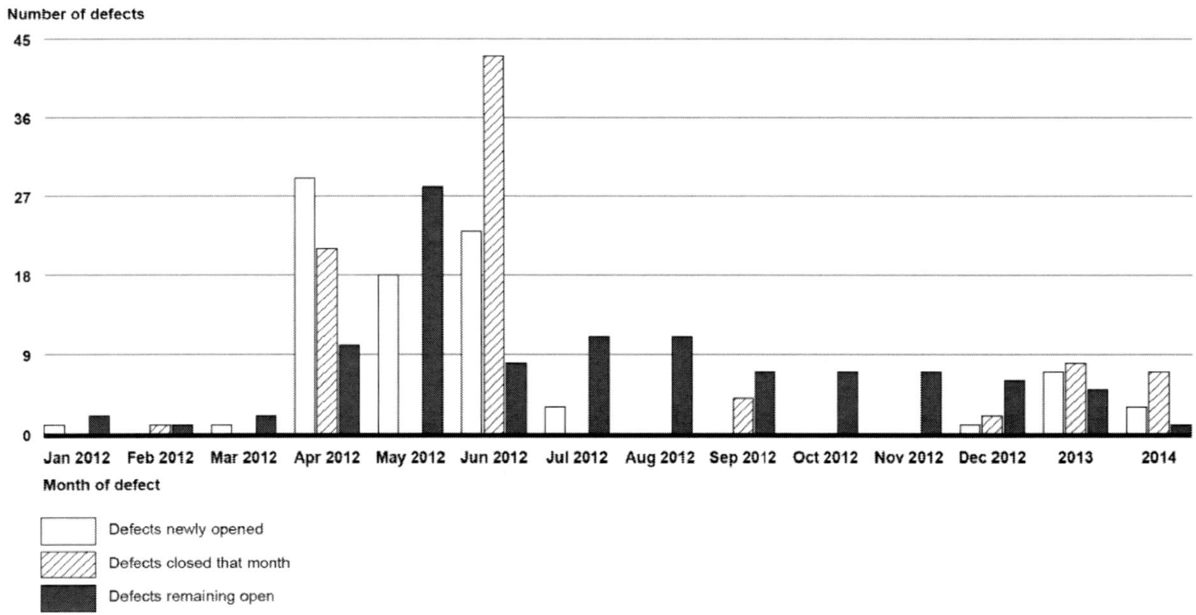

Source: GAO analysis of NOAA data. | GAO-15-60.

Figure 12. Number of Open and Closed Software Defects for the Geostationary Operational Environmental Satellite-R Program's Advanced Baseline Imager Instrument from 2012 to 2014.

While the backlog of hardware-related defects on the spacecraft remains high, NOAA has been effective in recent months in greatly reducing the number of software-related defects on the spacecraft. The total number of open software-related defects increased and remained high through February 2014, but then declined significantly in March and April 2014.

Defect Trends Show Decline in Defects for GLM Instrument

Defect data for the GLM instrument showed a decline in the total number of defects for both hardware and software components. While there were 117 unresolved hardware defects in May 2014, only 13 hardware defects remained unresolved as of September 2014. Trend data also showed a decline in the number of unresolved software defects. Specifically, the total number of software-related defects remaining open never increased above 7 over the period from March 2013 to January 2014. For half the year, there was no more than one open defect.

ABI Hardware and Software Defects Were Closed in Recent Months

Metrics for the ABI instrument show that all hardware defects and all but one software defect have been closed, likely because the first unit of the instrument was completed. The number of hardware defects occurring each month declined to near zero during the period February 2013 to April 2014. Less than ten ABI software defects were open at any point from September 2012 onward, and only 11 defects were newly opened during that time. Figure 11 shows the number of opened and closed ABI hardware defects by month, and figure 12 shows opened and closed defects for ABI software.

End Notes

[1] GAO, *High Risk Series: An Update*, GAO-13-283 (Washington, D.C.: February 2013).

[2] GAO, *GAO Cost Estimating and Assessment Guide*: *Best Practices for Developing and Managing Capital Program Costs*, GAO-09-3SP (Washington, D.C.: March 2009).

[3] See GAO, *Year 2000 Computing Crisis: Business Continuity and Contingency Planning*, GAO/AIMD-10.1.19 (Washington, D.C.: August 1998); National Institute of Standards and Technology, *Contingency Planning Guide for Federal Information Systems*, NIST 800-34

(Gaithersburg, Md.: May 2010); Software Engineering Institute, *CMMI® for Acquisition,* Version 1.3 (Pittsburgh, Pa.: November 2010).

[4] There are older GOES satellites in orbit, but they have been decommissioned.

[5] GAO, *Environmental Satellites: Focused Attention Needed to Improve Mitigation Strategies for Satellite Coverage Gaps,* GAO-13-865T, (Washington, D.C.: Sept. 19, 2013); *Geostationary Weather Satellites: Progress Made, but Weaknesses in Scheduling, Contingency Planning, and Communicating with Users Need to be Addressed,* GAO-13-597, (Washington, D.C.: Sept. 19, 2013); *Environmental Satellites: Focused Attention Needed to Mitigate Program Risks,* GAO-12-841T, (Washington, D.C.: June 27, 2012); *Geostationary Weather Satellites: Design Progress Made, but Schedule Uncertainty Needs to be Addressed*, GAO-12-576, (Washington, D.C.: June 26, 2012); *Geostationary Operational Environmental Satellites: Improvements Needed in Continuity Planning and Involvement of Key Users*, GAO-10-799 (Washington, D.C.: Sept. 1, 2010); *Geostationary Operational Environmental Satellites: Acquisition Has Increased Costs, Reduced Capabilities, and Delayed Schedules*, GAO-09-596T (Washington, D.C.: Apr. 23, 2009); *Geostationary Operational Environmental Satellites: Acquisition Is Under Way, but Improvements Needed in Management and Oversight*, GAO-09-323 (Washington, D.C.: Apr. 2, 2009); *Geostationary Operational Environmental Satellites: Further Actions Needed to Effectively Manage Risks*, GAO-08-183T (Washington, D.C.: Oct. 23, 2007); *Geostationary Operational Environmental Satellites: Progress Has Been Made, but Improvements Are Needed to Effectively Manage Risks*, GAO-08-18 (Washington, D.C.: Oct. 23, 2007);

[6] GAO-12-576.

[7] GAO-13-597.

[8] GAO-12-576.

[9] GAO-13-597.

[10] GAO, *2013 High-Risk Series: An Update*, GAO-13-359T (Washington, D.C.: Feb. 14, 2013).

[11] GAO-13-597.

[12] The program split the first of the five end-to-end tests into two increments, the first of which was completed in August 2014 and the last which is expected to be completed in December 2014.

[13] GAO-13-597.

[14] See, for example, GAO, *NASA: Assessments of Selected Large-Scale Projects,* GAO-11-239SP (Washington, D.C.: Mar. 3, 2011) and *NASA: Assessments of Selected Large-Scale Projects*, GAO-12-207SP (Washington, D.C.: Mar. 1, 2012).

[15] GAO, *Office of Personnel Management: Improvements Needed to Ensure Successful Retirement System Modernization*, GAO-08-345 (Washington, D.C.: Jan. 31, 2008) and *2000 Census: New Data Capture System Progress and Risks*, GAO/AIMD-00-61 (Washington, D.C.: Feb. 4, 2000).

[16] These components are costing more than their expected contract value; however, program officials expect that the total cost will remain within the program's contingency reserves.

[17] GAO-09-3SP.

[18] See, for example, GAO-11-239SP and GAO-12-207SP.

[19] GAO-09-3SP.

[20] Institute of Electrical and Electronics Engineers, *Standard for System and Software Verification and Validation*, IEEE Std 1012 (New York, N.Y.: May 25, 2012); Project Management Institute: A Guide to the Project Management Body of Knowledge (PMBOK® Guide), Fourth Edition, Project Management Institute, Inc. (PMI), 2008. Copyright and all

rights reserved.; Institute of Electrical and Electronics Engineers, *Software and systems engineering — Software testing*, ISO/TEC/IEEE Std 29119 (New York, N.Y.: Sept. 1, 2013); Institute of Electrical and Electronics Engineers, *Standard for Information Technology — Software life cycle processes — Implementation considerations*, IEEE/EIA Std 12207.2-1997 (New York, N.Y.: April 1998); Institute of Electrical and Electronics Engineers, *IEEE Standard for Software and System Test Documentation,* IEEE Std 829-2008 (New York, N.Y.: Jul. 10, 2008); Institute of Electrical and Electronics Engineers, *IEEE Standard Classification for Software Anomalies*, IEEE Std 1044-2009 (New York, N.Y.: Jan. 7, 2010); Software Engineering Institute, *CMMI® for Acquisition, Version 1.3* (Pittsburgh, Pa: November 2010); Software Engineering Institute, *CMMI® for Development, Version 1.3* (Pittsburgh, Pa: November 2010); GAO, *Year 2000 Computing Crisis: A Testing Guide*, GAO/AIMD-10.1.21 (Washington, D.C.: November 1998); *GAO Cost Estimating and Assessment Guide: Best Practices for Developing and Managing Capital Program Costs*, GAO-09-3SP (Washington, D.C.: March 2009); and *GAO Schedule Assessment Guide: Best Practices for Project Schedules (Exposure Draft)*, GAO-12-120G (Washington, D.C.: May 2012).

[21] See GAO, *Year 2000 Computing Crisis: Business Continuity and Contingency Planning*, GAO/AIMD-10.1.19 (Washington, D.C.: August 1998); National Institute of Standards and Technology, *Contingency Planning Guide for Federal Information Systems,* NIST 800-34 (Gaithersburg, Md.: May 2010); Software Engineering Institute, *CMMI® for Acquisition, Version 1.3* (Pittsburgh, Pa.: November 2010).

[22] GAO-13-597.

[23] While components are costing more than their expected contract value, program officials expect that the total cost will remain within the program's contingency reserves.

End Notes for Appendix I

[1] GAO-09-3SP.

[2] Institute of Electrical and Electronics Engineers, *Standard for System and Software Verification and Validation*, IEEE Std 1012 (New York, N.Y.: May 25, 2012); Project Management Institute: A Guide to the Project Management Body of Knowledge (PMBOK® Guide), Fourth Edition, Project Management Institute, Inc. (PMI), 2008. Copyright and all rights reserved.; Institute of Electrical and Electronics Engineers, *Software and systems engineering — Software testing*, ISO/TEC/IEEE Std 29119 (New York, N.Y.: Sept. 1, 2013); Institute of Electrical and Electronics Engineers, *Standard for Information Technology- Software life cycle processes – Implementation considerations*, IEEE/EIA Std 12207.2-1997 (New York, N.Y.: April 1998); Institute of Electrical and Electronics Engineers, *IEEE Standard for Software and System Test Documentation,* IEEE Std 829-2008 (New York, N.Y.: Jul. 10, 2008); Institute of Electrical and Electronics Engineers, *IEEE Standard Classification for Software Anomalies*, IEEE Std 1044-2009 (New York, N.Y.: Jan. 7, 2010); Software Engineering Institute, *CMMI® for Acquisition, Version 1.3* (Pittsburgh, Pa: November 2010); Software Engineering Institute, *CMMI® for Development, Version 1.3* (Pittsburgh, Pa: November 2010); GAO, *Year 2000 Computing Crisis: A Testing Guide*, GAO/AIMD-10.1.21 (Washington, D.C.: November 1998); *GAO Cost Estimating and Assessment Guide: Best Practices for Developing and Managing Capital Program Costs*, GAO-09-3SP (Washington, D.C.: March 2009); and *GAO Schedule*

Assessment Guide: Best Practices for Project Schedules (Exposure Draft), GAO-12-120G (Washington, D.C.: May 2012).

[3] The three components include the ABI instrument, the GLM instrument, and the ground system.

[4] See GAO, *Year 2000 Computing Crisis: Business Continuity and Contingency Planning*, GAO/AIMD-10.1.19 (Washington, D.C.: August 1998); National Institute of Standards and Technology, *Contingency Planning Guide for Federal Information Systems,* NIST 800-34 (Gaithersburg, Md.: May 2010); Software Engineering Institute, *CMMI® for Acquisition, Version 1.3* (Pittsburgh, Pa.: November 2010).

[5] GAO-13-597.

In: U.S. Weather Satellites
Editor: Louise O. Berkan

ISBN: 978-1-63482-726-3

Chapter 2

POLAR WEATHER SATELLITES: NOAA NEEDS TO PREPARE FOR NEAR-TERM DATA GAPS*

United States Government Accountability Office

WHY GAO DID THIS STUDY

NOAA established the JPSS program in 2010 to replace aging polar satellites and provide critical environmental data used in forecasting weather and measuring variations in climate. However, GAO and NOAA have previously reported that a gap in satellite data between the current satellite and the next one is likely. Given the criticality of satellite data to weather forecasting, the likelihood of a significant satellite data gap, and the potential impact of a gap on the health and safety of the U.S. population and economy, GAO added this issue to its High Risk List in 2013.

GAO was asked to review the JPSS program. GAO's objectives were to (1) evaluate NOAA's progress on the JPSS satellite program with respect to cost, schedule, and mitigation of key risks; (2) identify the benefits and challenges of alternatives for polar satellite gap mitigation; and (3) assess NOAA's efforts to establish and implement a comprehensive contingency plan for potential gaps in polar satellite data. To do so, GAO analyzed program

* This is an edited, reformatted and augmented version of a United States Government Accountability Office publication, No. GAO-15-47, dated December 2014.

management status reports, milestone reviews, and risk data; examined polar gap contingency plans; and interviewed experts as well as agency and contractor officials.

WHAT GAO RECOMMENDS

GAO is recommending NOAA track completion dates for risk mitigation activities, update its data gap assessment, address shortfalls in its contingency plan, prioritize mitigation projects most likely to address a gap, and report progress on all mitigation projects. NOAA concurred with GAO's recommendations and identified steps it is taking to implement them.

WHAT GAO FOUND

The Joint Polar Satellite System (JPSS) program has recently completed significant development activities and remains within its cost and schedule baselines; however, recent cost growth on key components is likely unsustainable and risks remain that could increase the potential for near-term satellite data gaps. For example, technical issues experienced while developing a key instrument have led to a very tight schedule. The National Oceanic and Atmospheric Administration (NOAA) is working to mitigate such risks, but is not tracking actual completion dates for its risk mitigation activities. In addition, while the program has reduced its estimate for a near-term satellite data gap in the afternoon orbit to only 3 months, its gap assessment was based on incomplete data (such as the increasing threat from space debris) and the agency has not updated its assessment to address these limitations. As shown below, a gap in satellite data may occur earlier and last longer than NOAA anticipates.

Experts within and outside of NOAA identified almost 40 alternatives for mitigating potential gaps in polar satellite data, which offer a variety of benefits and challenges. These alternatives include actions to prevent or limit a potential gap by providing JPSS-like capabilities, and actions that could reduce the impact of a potential gap by (a) extending and expanding the use of current data sources, (b) enhancing modeling and data assimilation,(c) developing new data sources, or (d) exploring opportunities with foreign and domestic partners.

However, obstacles to the alternatives, such as the time required to develop new instruments, may restrict them from being available to address a near-term gap.

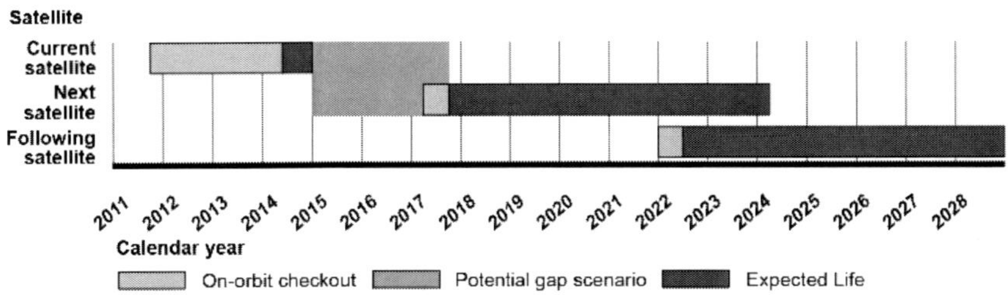

Source: GAO analysis based on NOAA and NASA data. | GAO-15-47.

Timeline for a Potential Gap in Polar Satellite Data in the Afternoon Orbit.

While multiple alternatives for mitigating a gap exist, NOAA's contingency plan focuses on a subset of these alternatives. NOAA has improved its contingency plan by identifying mitigation strategies and specific activities. However, the agency's plan has shortfalls such as not providing an assessment of available alternatives based on their cost and potential impacts. In addition, key projects affecting improvements to forecast models and assimilation of additional data sources have been delayed, but NOAA has not yet prioritized mitigation projects most likely to address a gap. Moreover, NOAA is not providing consistent or comprehensive reporting of its progress on all mitigation projects. Until NOAA addresses shortfalls in contingency planning, implements its most critical contingency activities before data gaps can occur in the near-term, and improves its progress monitoring, the agency will have less assurance that it is adequately prepared to deal with a gap in polar satellite coverage.

ABBREVIATIONS

A-DCS	Advanced Data Collection System
ATMS	Advanced Technology Microwave Sounder
CERES	Cloud and Earth's Radiant Energy System
CrIS	Cross-Track Infrared Sounder
DMSP	Defense Meteorological Satellite Program
DOD	Department of Defense

JPSS	Joint Polar Satellite System
Metop	Meteorological Operational (satellite)
NASA	National Aeronautics and Space Administration
NESDIS	National Environmental Satellite, Data, and Information Service
NOAA	National Oceanic and Atmospheric Administration
NPOESS	National Polar-orbiting Operational Environmental Satellite System
NWS	National Weather Service
S-NPP	Suomi National Polar-orbiting Partnership
OMPS	Ozone Mapping and Profiler Suite
POES	Polar-orbiting Operational Environmental Satellites
TSIS	Total and Spectral Solar Irradiance Sensor
VIIRS	Visible Infrared Imaging Radiometer Suite

* * *

December 16, 2014

The Honorable Lamar Smith
Chairman

The Honorable Eddie Bernice Johnson
Ranking Member
Committee on Science, Space, and Technology
House of Representatives

In 2010, when the Executive Office of the President's Office of Science and Technology Policy announced the decision to disband the acquisition of polar-orbiting environmental satellites through a tri-agency program, the National Oceanic and Atmospheric Administration (NOAA) initiated its own satellite program with assistance from the National Aeronautics and Space Administration (NASA).[1] This new program, called the Joint Polar Satellite System (JPSS), includes three satellites, the first of which was successfully launched in October 2011. These satellites are to provide critical environmental data that are used in weather forecasting and in measuring variations in climate.

Due in part to delays associated with disbanding the prior program and initiating the new one, NOAA officials acknowledge that there could be a gap in satellite data before the next satellite is launched and operational. Because of the criticality of satellite data to weather forecasting, the likelihood of a significant satellite data gap, and the potential impact of a gap on the health and safety of the U.S. population and economy, GAO added this issue to its High Risk List in 2013.[2]

Given your interest in the progress NOAA has made, you asked us to review the JPSS program and the potential for satellite data gaps. Our objectives were to (1) evaluate NOAA's progress on the JPSS satellite program with respect to cost, schedule, and mitigation of key risks; (2) identify the benefits and challenges of alternatives for polar satellite gap mitigation; and (3) assess NOAA's efforts to establish and implement a comprehensive contingency plan for potential gaps in polar satellite data.

To evaluate NOAA's progress on the JPSS satellite program, we analyzed changes in cost and schedule for key program components and milestones since the program established its baseline in July 2013. We examined mitigation plans and status reports for risks presented to NOAA's program management since July 2013. We also analyzed the program's latest assessment of the potential for satellite data gaps. We assessed the reliability of cost and schedule data by interviewing program officials to determine their process for conducting accuracy checks, by examining multiple program and project status reports at different points in time, and by comparing reported dates to source schedule data. We determined that the data were sufficiently reliable for our reporting purposes. We attended program management council meetings and interviewed relevant NOAA and NASA officials to discuss JPSS development status, the reasons for cost and schedule changes, risk mitigation processes and action plans, and the program's gap assessment for polar afternoon satellites. To identify the benefits and challenges of alternatives for polar satellite gap mitigation, we reviewed existing literature and developed a list of alternatives, interviewed experts from organizations within and outside of NOAA, and summarized the key benefits and challenges that experts commonly identified.

To assess NOAA's efforts to establish and implement a comprehensive contingency plan for potential gaps in polar satellite data, we analyzed the extent to which NOAA's February 2014 polar satellite contingency plan addressed shortfalls that we previously identified by comparing a prior version of NOAA's plan to government and industry best practices.[3] We also reviewed project plans, progress reports, and other relevant documentation to evaluate the extent to which NOAA is making progress against its contingency plan. We interviewed relevant NOAA officials regarding changes reflected in the

February 2014 contingency plan, oversight and reporting procedures, and implementation progress.

We conducted this performance audit from January 2014 to December 2014 in accordance with generally accepted government auditing standards. Those standards require that we plan and perform the audit to obtain sufficient, appropriate evidence to provide a reasonable basis for our findings and conclusions based on our audit objectives. Additional details on our objectives, scope, and methodology are provided in appendix I.

BACKGROUND

Since the 1960s, the United States has operated two separate operational polar-orbiting meteorological satellite systems: the Polar-orbiting Operational Environmental Satellite (POES) series, which is managed by NOAA, and the Defense Meteorological Satellite Program (DMSP), which is managed by the Air Force. These satellites obtain environmental data that are processed to provide graphical weather images and specialized weather products. These satellite data are also the predominant input to numerical weather prediction models, which are a primary tool for forecasting weather days in advance—including forecasting the path and intensity of hurricanes.[4] The weather products and models are used to predict the potential impact of severe weather so that communities and emergency managers can help prevent and mitigate its effects. Polar satellites also provide data used to monitor environmental phenomena, such as ozone depletion and drought conditions, as well as data sets that are used by researchers for a variety of studies such as climate monitoring.

Unlike geostationary satellites, which maintain a fixed position relative to the earth, polar-orbiting satellites constantly circle the earth in an almost north-south orbit, providing global coverage of conditions that affect the weather and climate. Each satellite makes about 14 orbits a day. As the earth rotates beneath it, each satellite views the entire earth's surface twice a day. Currently, a NOAA satellite (called the Suomi National Polar-orbiting Partnership, or S-NPP) and two operational DMSP satellites are positioned so that they cross the equator in the early morning, midmorning, and early afternoon. In addition, the government relies on a European satellite, called the Meteorological Operational (Metop) satellite, for satellite observations in the midmorning orbit.[5] In addition to the operational satellites, NOAA, the Air Force, and a European weather satellite organization maintain older satellites that still collect some data and are available to provide limited backup to the operational satellites should they degrade or fail. Figure 1 illustrates the current operational polar satellite constellation.

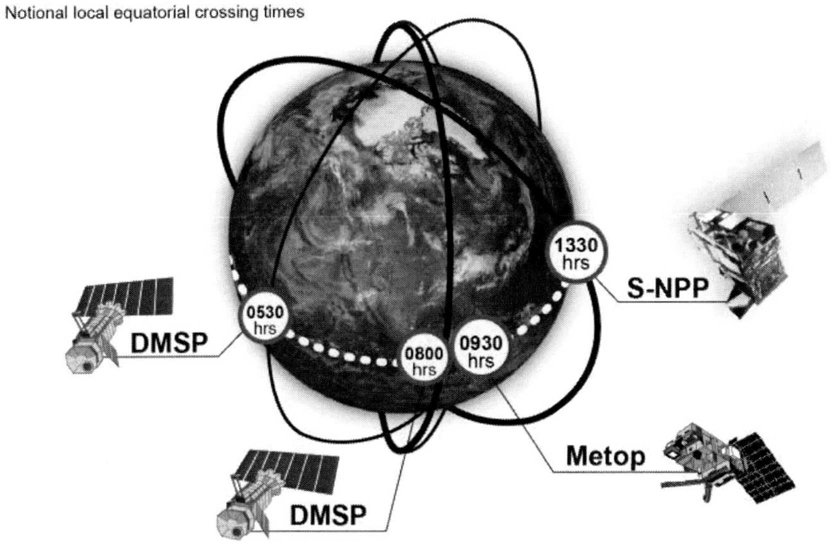

Sources: GAO based on NPOESS Integrated Program Office, NOAA, and DOD data; NASA/Goddard Space Flight Center Scientific Visualization Studio (earth); S-NPP image provided courtesy on University of Wisconsin-Madison Space Science and Engineering Center. | GAO-15-47.

Note: DMSP—Defense Meteorological Satellite Program, Metop—Meteorological Operational (satellite), and S-NPP—Suomi National Polar-orbiting Partnership.

Figure 1. Configuration of Operational Polar Satellites.

Polar Satellite Data and Products

Polar satellites gather a broad range of data that are transformed into a variety of products. Satellite sensors observe different bands of radiation wavelengths, called channels, which are used for remotely determining information about the earth's atmosphere, land surface, oceans, and the space environment. When first received, satellite data are considered raw data. To make them usable, processing centers format the data so that they are time-sequenced and include earth-location and calibration information. After formatting, these data are called raw data records. The centers further process these raw data records into channel-specific data sets, called sensor data records and temperature data records. These data records are then used to derive weather and climate products called environmental data records. These environmental data records include a wide range of atmospheric products detailing cloud coverage, temperature, humidity, and ozone distribution; land surface products showing snow cover, vegetation, and land use; ocean products depicting sea surface temperatures, sea ice, and wave height; and characterizations of the space environment. Combinations of these data records (raw, sensor, temperature, and environmental data records) are also used to derive more sophisticated products, including outputs from numerical weather models and assessments of climate trends. Figure 2 is a simplified depiction of the various stages of satellite data processing, and figure 3 depicts examples of two different environmental products.

The NPOESS Program: Inception, Challenges, and Termination

With the expectation that combining the POES and DMSP programs would reduce duplication and result in sizable cost savings, a May 1994 Presidential Decision Directive required NOAA and the Department of Defense (DOD) to converge the two satellite programs into a single one— the National Polar-orbiting Operational Environmental Satellite System (NPOESS)—that is capable of satisfying both civilian and military requirements.[6] The converged program, NPOESS, was considered critical to the nation's ability to maintain the continuity of data required for weather forecasting and global climate monitoring. NPOESS satellites were expected to replace the POES and DMSP satellites in the morning, midmorning, and afternoon orbits when they neared the end of their expected life spans.

To manage this program, DOD, NOAA, and the National Aeronautics and Space Administration (NASA) formed a tri-agency Integrated Program Office, with NOAA responsible for overall program management for the converged system and for satellite operations, the Air Force responsible for acquisition, and NASA responsible for facilitating the development and incorporation of new technologies into the converged system.

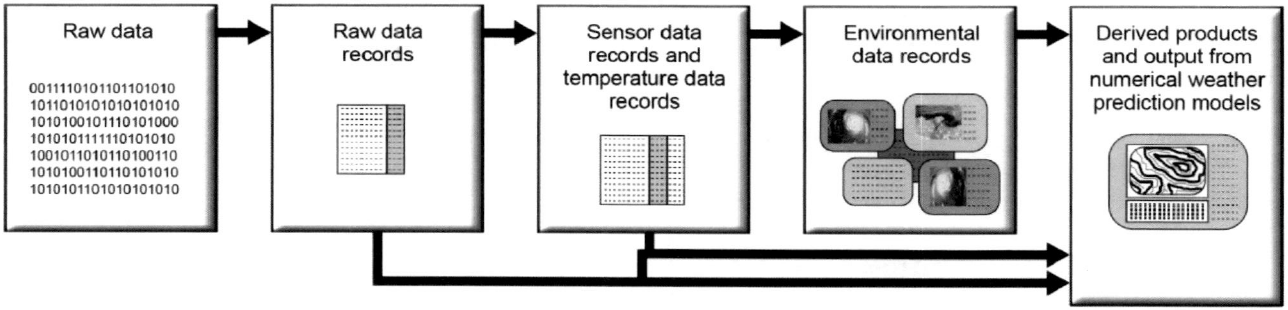

Figure 2. Stages of Satellite Data Processing.

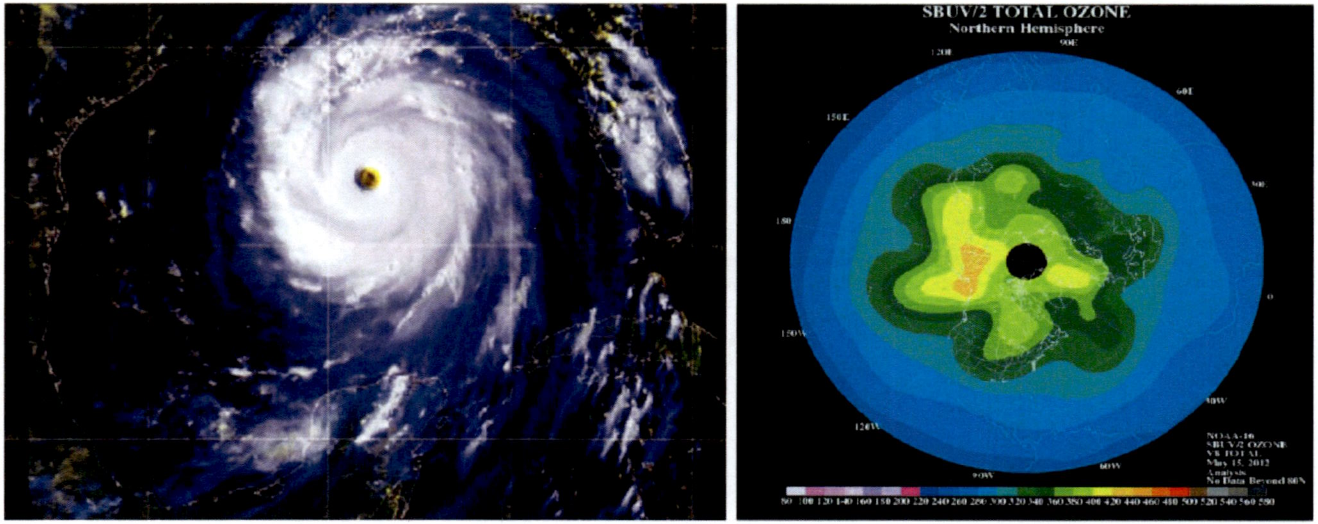

Figure 3. Examples of Environmental Products.

When the primary NPOESS contract was awarded in August 2002, the program was estimated to cost about $7 billion through 2018. It was to include the procurement and launch of 6 satellites over the life of the program, with each satellite hosting a subset of 13 instruments. The planned instruments included 11 environmental sensors and two systems supporting specific user services (see table 1). To reduce the risk involved in developing new technologies and to maintain climate data continuity, the program planned to launch the demonstration satellite in May 2006.[7] This satellite was intended to demonstrate the functionality of selected instruments that would later be included on the NPOESS satellites. The first NPOESS satellite was to be available for launch in March 2008.

Table 1. Anticipated National Polar-orbiting Operational Environmental Satellite System (NPOESS) Instruments, as of July 2002

Instrument	Instrument type	Description
Advanced technology microwave sounder (ATMS)	Environmental sensor	Measures microwave energy released and scattered by the atmosphere; to be used in combination with the cross-track infrared sounder to produce daily global atmospheric temperature, humidity, and pressure profiles.
Aerosol polarimetry sensor	Environmental sensor	Retrieves specific aerosol (liquid droplets or solid particles suspended in the atmosphere, such as sea spray, smog, and smoke) and cloud measurements.
Conical microwave imager/sounder	Environmental sensor	Collects microwave images and data needed to measure rain rate, ocean surface wind speed and direction, amount of water in the clouds, and soil moisture, as well as temperature and humidity at different atmospheric levels.
Cross-track infrared sounder (CrIS)	Environmental sensor	Collects measurements of the infrared radiation emitted and scattered by the Earth and atmosphere to determine the vertical distribution of temperature, moisture, and pressure in the atmosphere.
Data collection system	System providing services to users	Collects environmental data from platforms around the world and delivers them to users worldwide.

Table 1. (Continued)

Instrument	Instrument type	Description
Earth radiation budget sensor	Environmental sensor	Measures solar short-wave radiation and long-wave radiation released by the Earth back into space on a worldwide scale to enhance long-term climate studies.
Global positioning system occultation sensor	Environmental sensor	Measures the refraction of radio wave signals from the Global Positioning System and Russia's Global Navigation Satellite System to characterize the ionosphere and information related to the vertical distribution of temperature and moisture of the atmosphere.
Ozone	Environmental	Collects data needed to measure the amount and distribution of ozone in the Earth's
mapper/profiler suite (OMPS)	sensor	atmosphere. Consists of two components (nadir and limb) that can be provided separately.
Radar altimeter	Environmental sensor	Measures variances in sea surface height/topography and ocean surface roughness, which are used to determine sea surface height, significant wave height, and ocean surface wind speed and to provide critical inputs to ocean forecasting and climate prediction models.
Search and rescue	System	A subsystem that detects and locates aviators, mariners, and land-based users in distress.
satellite-aided tracking (SARSAT) system	providing services to users	
Space environmental sensor suite	Environmental sensor	Collects data to identify, reduce, and predict the effects of space weather on technological systems, including satellites and radio links.
Total and spectral solar irradiance	Environmental sensor	Monitors and captures total and spectral solar irradiance data.
sensor (TSIS)		

Instrument	Instrument type	Description
Visible infrared imaging radiometer suite (VIIRS)	Environmental sensor	Collects images and radiometric data used to provide information on the Earth's clouds, atmosphere, ocean, and land surfaces.

Source: GAO analysis of data from the former NPOESS Integrated Program Office. | GAO-15-47.

In the years after the program was initiated, NPOESS encountered significant technical challenges in sensor development, program cost growth, and schedule delays. By November 2005, we estimated that the program's cost had grown to $10 billion, and the schedule for the first launch was delayed by almost 2 years.[8] These issues led to a 2006 decision to restructure the program, which reduced the program's functionality by decreasing the number of planned satellites from 6 to 4, and the number of instruments from 13 to 9. As part of the decision, officials decided to reduce the number of orbits from three (early morning, midmorning, and afternoon) to two (early morning and afternoon) and to rely solely on the European satellites for midmorning orbit data.

Even after the restructuring, however, the program continued to encounter technical issues in developing two sensors, significant triagency management challenges, schedule delays, and further cost increases. Because the schedule delays could lead to satellite data gaps, in March 2009, agency executives decided to use S-NPP as an operational satellite. [9] Later, faced with costs that were expected to reach about $15 billion and launch schedules that were delayed by over 5 years, the Executive Office of the President announced that NOAA and DOD would no longer jointly procure the NPOESS satellite system; instead, each agency would plan and acquire its own satellite system.[10] Specifically, NOAA would be responsible for the afternoon orbit and the observations planned for the first and third satellites and DOD would be responsible for the early morning orbit and the observations planned for the second and fourth satellites. The partnership with the European satellite agencies for the midmorning orbit was to continue as planned.

When this decision was announced, NOAA and NASA immediately began planning for a new satellite program in the afternoon orbit called JPSS. DOD began planning for a new satellite program in the morning orbit, called the Defense Weather Satellite System, but later decided to terminate the program and reassess its requirements, as directed by Congress.

Overview of the JPSS Program

After the decision was made to disband the NPOESS program in 2010, NOAA began the JPSS satellite program. Key plans included:

- relying on NASA for system acquisition, engineering, and integration;
- completing, launching, and supporting S-NPP;
- acquiring and launching two satellites for the afternoon orbit, called JPSS-1 and JPSS-2;
- developing and integrating five sensors on the two satellites;
- finding alternative host satellites for selected instruments that would not be accommodated on the JPSS satellites, including the Total and Spectral Solar Irradiance Sensor (TSIS), the Advanced Data Collection System (A-DCS), and the Search and Rescue Satellite-Aided Tracking (SARSAT) system; and
- providing ground system support for S-NPP, JPSS, and the Defense Weather Satellite System; data communications for Metop and DMSP; and data processing for NOAA's use of microwave data from an international satellite.

In 2010, NOAA estimated that the life cycle costs of the JPSS program would be approximately $11.9 billion for a program lasting through fiscal year 2024, which included $2.9 billion in NOAA funds spent on NPOESS through fiscal year 2010.[11]

Subsequently, the agency undertook a cost estimating exercise where it validated that the cost of the full set of JPSS functions from fiscal year 2012 through fiscal year 2028 would be $11.3 billion.

After adding the agency's sunk costs, which had increased to $3.3 billion through fiscal year 2011, the program's life cycle cost estimate totaled $14.6 billion.[12] This amount was $2.7 billion higher than the $11.9 billion estimate for JPSS when NPOESS was disbanded in 2010.

Table 2. Comparison of the National Polar-orbiting Operational Environmental Satellite System (NPOESS) and the Joint Polar Satellite System (JPSS) Programs, at Different Points in Time

Key area	NPOESS after it was restructured (as of June 2006)	NPOESS prior to being disbanded (as of February 2010)	JPSS program (as of May 2010)	JPSS program (as of June 2012)	JPSS program (as of September 2013)
Life cycle	1995-2026	1995-2026	2010-2024	2010-2028	2010-2025
Estimated life cycle cost	$12.5 billion	$13.95+ billion[a]	$11.9 billion (which includes about $2.9 billion spent through fiscal year 2010 on NPOESS)	$12.9 billion (which includes about $3.3 billion spent through fiscal year 2011 on NPOESS and JPSS)	$11.3 billion (which includes about $4.3 billion spent through fiscal year 2012 on NPOESS and JPSS)
Number of satellites	4 (in addition to S-NPP)	4 (in addition to S-NPP)	2 (in addition to S-NPP)	2 (in addition to S-NPP)	2 (in addition to S-NPP)
Number of orbits	2 (early morning and afternoon; would rely on European satellites for midmorning orbit data)	2 (early morning and afternoon; would rely on European satellites for midmorning orbit data)	1 (afternoon orbit) (DOD and European satellites would provide early and midmorning orbits, respectively)	1 (afternoon orbit) (DOD and European satellites would provide early and midmorning orbits, respectively)	1 (afternoon orbit) (DOD and European satellites would provide early and midmorning orbits, respectively)
Launch schedule	S-NPP by Jan. 2010 First satellite (C1) by Jan. 2013 C2 by Jan. 2016 C3 by Jan. 2018 C4 by Jan. 2020	S-NPP no earlier than Sept. 2011 C1 by March 2014 C2 by May 2016 C3 by Jan. 2018 C4 by Jan. 2020	S-NPP—no earlier than Sept. 2011 JPSS-1 available in 2015 JPSS-2 available in 2018	S-NPP—successfully launched in Oct. 2011 JPSS-1 by March 2017 JPSS-2 by Dec. 2022	S-NPP—successfully launched in Oct. 2011 JPSS-1 by March 2017 JPSS-2 by Dec. 2021

Table 2. (Continued)

Key area	NPOESS after it was restructured (as of June 2006)	NPOESS prior to being disbanded (as of February 2010)	JPSS program (as of May 2010)	JPSS program (as of June 2012)	JPSS program (as of September 2013)
Life cycle	1995-2026	1995-2026	2010-2024	2010-2028	2010-2025
Number of sensors	S-NPP: 4 sensors C1: 6 sensors C2: 2 sensors 6 sensors 2 sensors	S-NPP: 5 sensors C1: 7 sensors[b] C2: 2 sensors 6 sensors 2 sensors	S-NPP: 5 sensors JPSS-1 5 sensors[c] JPSS-2: 5 sensors	S-NPP: 5 sensors JPSS-1: 5 sensors JPSS-2: 5 sensors Free flyer-1 and-2: 1 sensor and 2 user services systems[d]	S-NPP: 5 sensors JPSS-1: 5 sensors JPSS-2: 5 sensors[e] No free flyers[f]

Source: GAO analysis of NOAA, DOD, and task force data. | GAO-15-47

[a]Although the program baseline was $13.95 billion in February 2010, we estimated in June 2009 that this cost could grow by about $1 billion. In addition, officials from the Executive Office of the President stated that they reviewed life cycle cost estimates from DOD and the NPOESS program office of $15.1 billion and $16.45 billion, respectively.

[b]In May 2008, the NPOESS Executive Committee approved an additional sensor—Total and Spectral Solar Irradiance Sensor—for the C1 satellite.

[c]The five sensors are the Advanced Technology Microwave Sounder, Clouds and the Earth's Radiant Energy System (CERES), Cross-Track Infrared Sounder, Ozone Mapping and Profiler Suite, and Visible Infrared Imaging Radiometer Suite. NOAA committed to finding an alternative spacecraft and launch accommodation for the Total and Spectral Solar Irradiance Sensor, the Advanced Data Collection System, and the Search and Rescue Satellite-Aided Tracking system.

[d]NOAA planned to launch two stand-alone satellites, called free flyer satellites, to accommodate the Total and Spectral Solar Irradiance Sensor, the Advanced Data Collection System, and the Search and Rescue Satellite-Aided Tracking system.

[e]In its fiscal year 2014 budget request, NOAA transferred responsibility for two sensors to NASA—the Radiation Budget Instrument (formerly known as CERES) and OMPS-L and plans to accommodate these sensors on the JPSS-2 satellite as long as they do not impact the likelihood of mission success.

[f]NOAA canceled Free flyer-1 and established Free flyer-2 as a new program outside the JPSS program. This new program, called the Solar Irradiance, Data, and Rescue (SIDAR) mission, is to accommodate the Total and Spectral Solar Irradiance Sensor, the Advanced Data Collection System, and the Search and Rescue Satellite-Aided Tracking system.

In working with the Office of Management and Budget to establish the President's fiscal year 2013 budget request, NOAA officials agreed to cap the JPSS life cycle cost at $12.9 billion through 2028, to fund JPSS at roughly $900 million per year through 2017, and to merge funding for two climate sensors into the JPSS budget. Because this cap was $1.7 billion below the expected $14.6 billion life cycle cost of the full program, by October 2012, NOAA decided to remove selected elements from the satellite program, such as the number of ground-based receptor stations (thus affecting the time that it takes for products to reach end users) and the number of interface data processing segments.

Subsequently, according to NOAA officials, the administration directed the agency to begin implementing additional changes in the program's scope and objectives in order to meet the agency's highest-priority needs for weather forecasting and reduce program costs from $12.9 billion to $11.3 billion. By April 2013, the program had decided to, among other things, transfer requirements for certain climate sensors to NASA, cancel one of two planned free-flyer missions, and transfer the remaining free-flyer mission to a new program within NOAA called the Solar Irradiance, Data, and Rescue (SIDAR) mission.[13] The program also reduced program costs by removing funding for 3 years of operations at the end of the JPSS mission; the program is now funded through 2025 even though the JPSS-2 satellite is expected to be operational until 2028.[14]

Table 2 compares the planned cost, schedule, and scope of NOAA's satellite programs at different points in time.

Recent GAO Work Recommended Actions to Address Shortfalls in Establishing a Contingency Plan for Potential Polar Satellite Data Gaps

We previously reported that the polar satellite constellation—consisting of DOD, European, and NOAA satellites in the early morning, midmorning, and afternoon polar orbits—was becoming increasingly unreliable and that satellite data gaps in the morning or afternoon polar orbits would lead to less accurate and timely weather forecasting.[15] As a result, advanced warning of extreme events would be affected. Such extreme events could include hurricanes, storm surges, and floods. For example, the National Weather Service performed case studies to demonstrate how its forecasts would have been affected if there were no polar satellite data in the afternoon orbit, and noted that its forecasts for the "Snowmaggedon" winter storm that hit the Mid-Atlantic coast in February 2010 would have predicted a less intense storm further east, with about half of the precipitation at 3, 4, and 5 days before the event. Specifically, the models would have under-forecasted the amount of snow by at least 10 inches. Similarly, a European weather organization recently reported that forecasts of Hurricane Sandy's track could have been hundreds of miles off without polar-orbiting satellites—rather than identifying the New Jersey landfall within 30 miles 4 days before landfall, the models would have shown the storm remaining at sea.[16]

In June 2012, we reported that the length of an afternoon polar satellite data gap could span from 17 months to 3 years or more under different scenarios.[17] Because the degradation in forecasts and warnings resulting from gaps in polar weather satellite data would place lives, property, and our nation's critical infrastructure in danger, we added the potential gap in weather satellite data to our biennial High-Risk list in February 2013.[18] In that report, we noted that, while NOAA had established plans to address potential gaps in polar satellite data, those plans were only the beginning. We noted that NOAA must make difficult decisions about (1) whether and how to extend support for legacy satellite systems so that their data might be available if needed, (2) how much time and resources to invest in improving satellite models so that they assimilate data from alternative sources, (3) whether to pursue international agreements for access to additional satellite systems and how best to resolve any security issues with the foreign data, (4) when and how to test the value and integration of alternative data sources, and (5) how these preliminary mitigation plans will be integrated with the agency's broader end-to-end plans for sustaining weather forecasting capabilities.

More recently, in September 2013 we reported that NOAA had identified multiple ways to help mitigate expected gaps in polar satellite data, but it had not yet developed and implemented a comprehensive contingency plan and its current mitigation plan had shortfalls when compared to government and industry best practices.[19] We recommended that NOAA establish a comprehensive contingency plan for potential satellite data gaps in the polar orbit that is consistent with contingency planning best practices. NOAA agreed with our recommendation and identified steps it planned to take to update its plan through July 2014.

JPSS is Meeting Cost and Schedule Baselines and Working to Mitigate Key Risks, but Potential Data Gaps Could Occur Sooner and Last Longer Than Expected

While the JPSS program has recently completed significant development activities and remains within its cost and schedule baselines, risks remain that could increase the potential for near-term satellite data gaps in the afternoon orbit. For example, technical issues experienced during instrument development have led to reduced schedule margin and the ground system may not be ready in time for the next satellite launch. NOAA is working to mitigate these risks, but has not identified completion dates for key activities. Also, while the program has reduced its estimate for a near-term satellite data gap in the afternoon orbit, the assessment did not consider the potentially life-limiting impacts to S-NPP and other polar satellites due to an increasing threat from space debris. As a result, a gap in satellite data may occur earlier and last longer than NOAA anticipates.

NOAA is Developing JPSS within Cost and Schedule Baselines, but Key Components Have Experienced Cost Growth and Milestone Delays

NOAA is currently developing JPSS within its cost and schedule baselines. In July 2013, NOAA formally established a cost baseline of $11.3 billion for the program through fiscal year 2025, launch dates of no later than March 2017 for JPSS-1 and December 2021 for JPSS-2, and completion dates for other important development milestones, such as the JPSS-1 critical design review and satellite integration and test review.

The JPSS cost baseline remains at $11.3 billion. NOAA currently estimates that the program's development, maintenance, and operations will cost $10.6 billion, meaning that it has about $700 million in reserve funding for unanticipated issues when compared to its cost baseline. In addition, the launch dates for the final two JPSS satellites remain on schedule since NOAA established the program's schedule baseline. JPSS-1 is still expected to launch no later than March 2017 and JPSS-2 is still expected to launch no later than December 2021. In addition, the program completed a major development milestone—the critical design review for the JPSS-1 mission—in April 2014.

This is a significant accomplishment because the review affirms that the satellite design is appropriately mature to continue with development, and that the program is on track to meet mission requirements with appropriate margins and acceptable risk.

Key JPSS Components Have Experienced Cost Growth

While JPSS development is within its overall life cycle cost baseline, key JPSS components have experienced cost growth. Since July 2013, the total program cost estimate has increased by two percent, or $222 million. More than half of this increase is for three instruments. Specifically, since July 2013, the ATMS instrument's cost increased by 17 percent of its total, the CrIS instrument's cost estimate increased by 23 percent of its total, and the VIIRS instrument's cost estimate increased by 6 percent of its total. Table 3 compares cost estimates for JPSS program components since the program's key milestone review in July 2013.

Table 3. Changes in Cost Estimates for Joint Polar Satellite System Components between July 2013 and July 2014

JPSS program components	Program estimate ($M), as of July 2013	Program estimate ($M), as of July 2014	Difference ($M)	Percentage change
Flight segment	2,758	2,983	225	8%
Ground segment	1,318	1,274	-44	-3%
Program office (includes satellite operations and sustainment)	3,460	3,501	41	1%
Total:				
Total cost to complete (FY10-25)	7,537	7,759	222	3%
Legacy (enacted)	2,848	2,848	0	0%
Life cycle cost	10,385	10,607	222	2%

Source: GAO analysis of NOAA data. | GAO-15-47

Program officials cited multiple reasons for these cost increases, including technical issues identified with the CrIS instrument during environmental testing, the need for additional environmental testing dry runs and calibration effort for VIIRS, and the purchase of critical spare parts that are expected to help accelerate instrument development on the JPSS-2 satellite. While a 2 percent total cost increase for the program since July 2013 may appear to be relatively small given remaining reserves of roughly $700 million, the program cannot sustain this rate of growth over the long term. If JPSS costs were to continue to grow at the same rate of 2 percent of its life cycle cost estimate annually, the program could surpass its cost baseline by 2018 and end up costing $2 billion more than expected by 2025. Therefore, moving forward, it will be important for NOAA and NASA managers to aggressively monitor and control components that are threatening to exceed their expected costs.

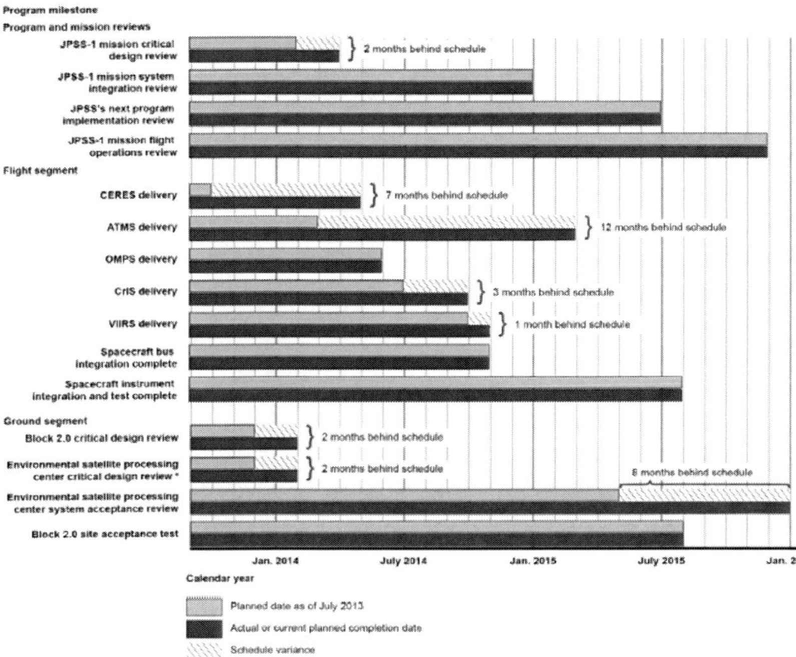

Source: GAO analysis of NOAA data. | GAO-15-47.

Note: CERES—Clouds and Earth's Radiant Energy System, ATMS—Advanced Technology Microwave Sounder, OMPS—Ozone Mapping and Profiler Suite, CrIS—Cross-Track Infrared Sounder, and VIIRS—Visible Infrared Imaging Radiometer Suite.

[a]The environmental satellite processing center critical design review milestone was for JPSS-supported components only.

Figure 4. Changes in Selected Joint Polar Satellite System (JPSS) Milestones since July 2013.

Key JPSS Components Have Experienced Milestone Delays

While the JPSS-1 launch date has not yet been affected, key components have encountered milestone delays in development and testing. For example, since July 2013, the delivery of the ATMS instrument slipped by 12 months and the system acceptance review for the environmental satellite processing center slipped by 8 months. Figure 4 compares planned completion dates from the program's key milestone review in July 2013 with actual or current planned completion dates for selected milestones.

Program officials provided multiple reasons for the schedule changes, including technical issues the ATMS instrument experienced during thermal vacuum testing, a schedule adjustment to align the environmental processing center site acceptance review with NOAA's geostationary satellite acquisition, and the October 2013 government shutdown. Officials explained that the government shutdown led the program to delay critical design reviews for ground system subcomponents that required government managers' participation, which subsequently caused delays to completion of the overall ground system critical design and the JPSS-1 mission critical design review.

These delays have caused a reduction in schedule margin prior to the JPSS-1 satellite integration and testing phase. Specifically, because of the technical issues experienced by ATMS, the instrument has now become the critical path for the entire JPSS-1 mission and only 1 month of schedule reserve remains until its expected delivery in March 2015.[20] While the program reports 6 months of schedule reserve remaining for the JPSS-1 mission through its launch readiness date, which exceeds NASA Goddard Space Flight Center development standards, with ATMS on the critical path, it will be important for NOAA and NASA managers to quickly resolve the instrument's technical issues before it becomes a more serious threat to the mission schedule and launch date.

JPSS is Addressing Key Risks, but Does Not Track Actual Completion Dates for Important Mitigation Activities

The JPSS program's risk management guidance calls for identifying risks and developing action plans for addressing the risks. These action plans are to include a list of steps to mitigate the risks and when those steps are to be completed. Further, sound risk management practices call for tracking actual completion dates against the planned schedule when implementing risk mitigation plans.[21]

Since its key milestone review in July 2013, JPSS has identified approximately 40 key risks and has presented these during NOAA monthly program management council reviews. To mitigate these risks, the program has developed mitigation plans and started implementing those plans. Three key risks and efforts to mitigate them include:

- **JPSS-1 spacecraft vibration impacts:** Early tests showed that the spacecraft may not meet requirements associated with vibration testing, which could affect the instruments' performance. Program officials are taking steps to mitigate this risk, including working with instrument contractors to perform vibration analysis and testing and to address instances where performance does not meet specification.
- **Space debris environment:** NASA recently updated its assessment of the amount of orbital debris in the space environment to include expected increases in satellite launches and collisions between debris particles. Based on its most recent analysis, NASA identified an increased likelihood of a collision with space debris at the altitude where JPSS satellites operate. A collision with space debris could cause catastrophic damage to the satellites (see appendix II for additional information on the risks from space debris). NOAA and NASA regularly maneuver their spacecraft to avoid debris tracked by DOD's Space Surveillance Network (typically larger than 5 to 10 centimeters in size for the JPSS orbit); however, it is more difficult to address debris smaller than 5 centimeters. The JPSS program is taking steps to mitigate the risk for JPSS-1 from debris that is less than 2 centimeters, such as using shielding to protect propellant tanks, batteries, and other critical components. However, there is no shielding on S-NPP, which is currently in orbit. Further, the JPSS program currently has no way to protect the satellites from debris between 2 and 5 centimeters, as discussed later in this report.
- **ATMS amplifier issues may degrade instrument performance:** Problems with the ATMS instrument's amplifiers have been affecting the instrument's channels. NASA managers expressed a concern that the instrument could experience degraded performance in orbit. The program has traced several of the channel issues to intermediate frequency amplifiers that have been shown to have foreign object debris. In June 2014, the program initiated a mitigation plan to remove, test, clean, and rework all intermediate frequency amplifiers in the instrument.

While the program is developing mitigation plans and monitoring the execution of those plans for key risks, it has not tracked actual completion dates for the steps it is taking to address its key risks. Thus, it is not clear whether mitigation activities are being completed on, behind, or ahead of schedule. The program is not tracking actual completion dates because the program's risk management guidance does not require such tracking. While this is not required by the program's risk guidance, leading organizations including the Software Engineering Institute and Project Management Institute consider it a best practice because it allows for effective oversight. Until it tracks actual completion dates for key risk mitigation activities, NOAA will have less assurance that key risks that could affect JPSS's cost, launch schedule, and performance in orbit are being effectively mitigated.

NOAA Anticipates a 3-month Satellite Data Gap, but Other Scenarios Show Gaps that Occur Sooner and Last Longer than Expected

NOAA currently anticipates a shorter gap in satellite data than it previously estimated. In October 2013, the JPSS program office reported that a gap in between the S-NPP satellite and the JPSS-1 satellite could be as short as 3 months, which is 15 months less than NOAA estimated in 2012. This reduction is primarily due to factoring in a longer life for S-NPP because of its relatively strong performance since its launch, removing previous assumptions about satellite mechanisms (such as seals and bearings) wearing out, and reducing the expected length of the JPSS-1 satellite's on-orbit checkout phase. The program also reported that less-capable legacy satellites could provide backup data through 2019 to mitigate potential launch problems or premature instrument failures on the JPSS-1 satellite.

While it is encouraging that the program has concluded that a potential gap between S-NPP and JPSS-1 could be less than previously estimated, there are several reasons why this potential gap could occur sooner and last longer than NOAA currently anticipates.

- **Inconsistent launch date plans:** The program's analysis that JPSS-1 will be operational by June 2017 is inconsistent with NOAA's launch date commitment and the program's estimated checkout and calibration/validation period. Given that NOAA's external launch date commitment is no later than March 2017 and the program office estimates 6 months for on-orbit checkout and calibration/validation before the satellite data are operational, JPSS-1 would not be available until September 2017 or later, which increases the length of the potential gap from a minimum of 3 months to a minimum of 6 months.

- **Unproven predictions about the on-orbit checkout and validation phase:** The on-orbit checkout and calibration/validation phase could take longer than the program's estimated 6 months if there are issues with the instruments or ground systems. As a precedent, it has taken the JPSS program more than 2 years to fully validate the highest priority data products from the S-NPP satellite.[22] In addition, if the program office's algorithm development team does not incorporate certain pre-launch updates into the ground system, additional algorithm work may be needed after the satellite launches, which could extend the validation time frame. Program officials noted that it took longer to validate data for S-NPP because it was the first satellite of the JPSS series. Program officials also stated that they are incorporating lessons learned from S-NPP's calibration and validation, such as improving communication between algorithm teams and operational users, and expect that algorithm work completed for S-NPP can be reused for JPSS-1 to shorten its validation phase.
- **Exclusion of a key risk:** The program's gap assessment does not factor in the potential for satellite failures from space debris that are too small to be tracked and avoided. As an example of the potential impact that small debris can have on JPSS satellites, an initial risk assessment of JPSS-1's propellant tanks indicated that, without shielding, there was a 29 percent risk of failure over the mission's life time, which could be reduced to as little as 0.5 percent risk of failure with shielding. However, S-NPP does not have shielding to protect it from debris smaller than 2 centimeters. Moreover, like all satellites in this orbit, even if S-NPP had shielding, the program would have limited options on how to protect the satellite from debris between 2 and 5 centimeters. Thus, the S-NPP mission could end earlier than its 5-year design life, resulting in a gap period that occurs sooner and lasts longer than expected.

NOAA officials acknowledge that the gap assessment has several limitations and stated that they plan to update it. While it is a sound management practice to identify time frames for achieving objectives, the agency has not yet established a time frame for updating its assessment.[23] Until NOAA updates its gap assessment to include more accurate assumptions and key risks, such as the potential effect of space debris, the agency risks making decisions based on a limited understanding of the potential timing and length of a gap.

Alternative Scenarios Depict a Longer Gap

While NOAA believes the gap between the time when the S-NPP satellite reaches the end of its life and the time when the JPSS-1 satellite is to be in orbit and operational could be as little as 3 months, we believe that there are other plausible scenarios where a gap could be substantially longer. Depending on different assumptions, a gap could span from 11 months to more than 3 years.

In one scenario, S-NPP would last its full expected 5-year life (to October 2016), and JPSS-1 would launch as soon as possible (in March 2017) and undergo calibration/validation for 6 months as predicted by the JPSS program office (until September 2017). In that case, the data gap would extend 11 months.

In another scenario, S-NPP could fail this year as a result of impacts to the satellite from orbital debris. Assuming that the JPSS-1 launch occurred in March 2017 and the satellite data was certified for official use by September 2017, this gap would extend for 33 months. Of course, any problems with JPSS-1 development resulting in launch delays, launch problems, or delays in the planned 6-month calibration/validation period could extend the gap period. For example, another scenario could occur in which S-NPP lasts until October 2016 and JPSS-1 experiences launch problems or early instrument failures, resulting in a gap until JPSS-2 launches (in December 2021) and undergoes calibration/validation for at least 6 months (until June 2022). In that scenario, there could be a gap of 5 years and 8 months. Figure 5 depicts possible gap scenarios.

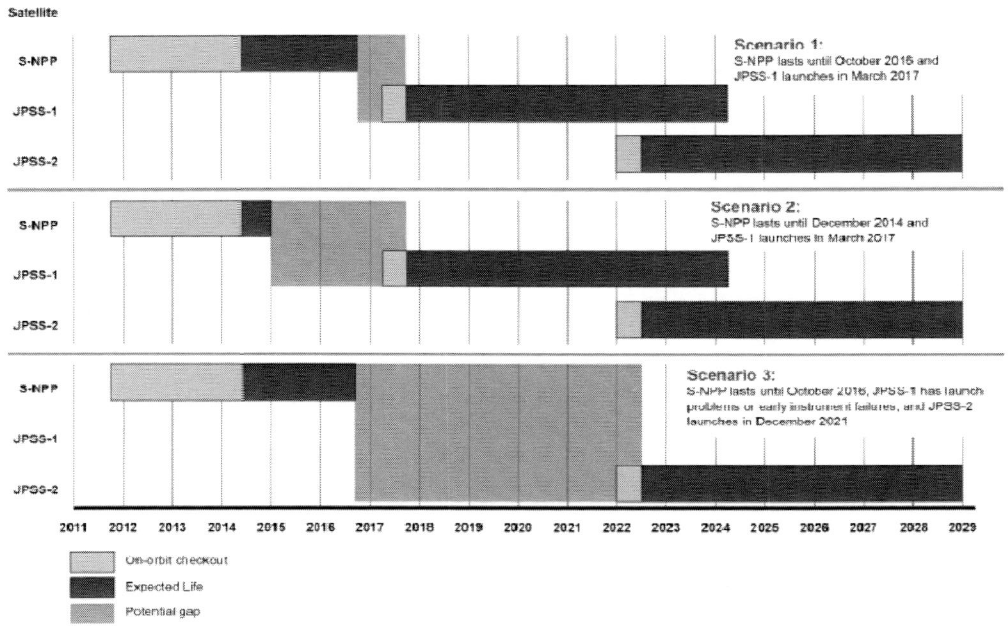

Source: GAO analysis based on NOAA and NASA data. | GAO-15-47.

Note: S-NPP—Suomi National Polar-orbiting Partnership, JPSS-1—Joint Polar Satellite System-1, and JPSS-2—Joint Polar Satellite System-2.

Figure 5. Potential Scenarios for a Gap in Polar Satellite Data in the Afternoon Orbit.

Multiple Alternatives Exist for Mitigating a Satellite Data Gap— Each with Benefits and Challenges—but They are Unlikely to Fully Address a Near-term Gap

NOAA and other experts in the weather satellite and modeling community have identified multiple alternatives for mitigating a potential gap in polar satellite data in the afternoon orbit. The alternatives can be separated into two general categories. The first category includes actions to prevent or limit a potential gap by providing JPSS-like capabilities. The second category includes actions that could reduce the impact of a potential gap by (a) extending and expanding the use of current data sources with capabilities similar to the JPSS program; (b) enhancing modeling and data assimilation; (c) developing new data sources; or (d) exploring opportunities with foreign and domestic partners. See table 4 for examples of alternatives within each of the categories.

Table 4. Key Alternatives for Mitigating a Potential Satellite Data Gap in the Afternoon Orbit

Gap mitigation categories andsubcategories	Key alternatives
Prevent or limit a potential gap	• extend S-NPP mission life • ensure the timely and successful JPSS-1 launch • accelerate JPSS-2 launch readiness • establish a "gap filler" mission • establish the SIDAR mission
Reduce the impact of a potential gap	
• Extend and expand use of current data sources	• extend legacy satellites (POES, Aqua) • rely on DOD DMSP • rely on European Metop satellites • expand aircraft observations from domestic and international commercial flights • expand Global Positioning System radio occultation data • expand the use of direct readout imagery from satellites over Alaska • rely on NASA's Aura satellite in the afternoon orbit • rely on NOAA's geostationary satellites for data collection

Gap mitigation categories andsubcategories	Key alternatives
• Enhance modeling and data assimilation	• enhance 4-dimensional data assimilation • accelerate global model research to operations • increase high-performance computing capacity for research and operations • enhance the blending of global models • add additional direct readout sites to reduce data latency
• Develop new sources	• develop geostationary satellite imagery-derived moisture profiles • develop atmospheric motion vectors (imagery-derived wind information) • develop cloud-impacted infrared and microwave soundings • develop doppler wind lidar • develop products from NASA's Global Precipitation Measurement mission • implement targeted observations from unmanned aircraft
• Explore partnership opportunities	• explore accelerated procurement (sensors, satellites, and launch vehicles) • explore contracting for small satellites provided by industry • explore commercial data buys • explore hosted payloads • explore prototype-stage technologies • explore the use of data from China's Feng-Yun 3 series • explore the use of Canada's proposed Polar Communications and Weather mission • explore the use of European Sentinel satellites • explore the use of Japan's Global Change Observation Mission – Water 2 and Water 3 satellites • explore the use of Japan's Global Change Observation Mission for Climate satellite • explore the use of International Space Station as a host for instruments • explore the use of a French and Indian satellite in the early morning orbit

Table 4. (Continued)

Gap mitigation categories andsubcategories	Key alternatives
	• explore the use of Russian search and rescue capability in the early morning orbit • explore the use of proposed Canadian small satellite as SARSAT "gap filler"

Source: GAO analysis based on interviews with experts from weather satellite-related organizations. | GAO-15-47

Note: DOD—Department of Defense; DMSP—Defense Meteorological Satellite Program; JPSS-1— Joint Polar Satellite System-1; JPSS-2—Joint Polar Satellite System-2; Metop—Meteorological Operational (satellite); NASA—National Aeronautics and Space Administration; POES—Polar-orbiting Operational Environmental Satellites; NOAA—National Oceanic and Atmospheric Administration; SARSAT—Search and Rescue Satellite-Aided Tracking system; SIDAR—Solar Irradiance, Data, and Rescue; and S-NPP—Suomi National Polar-orbiting Partnership.

Alternatives for Preventing or Limiting a Satellite Data Gap Offer Benefits and Pose Challenges

There are several ways to reduce the likelihood of a potential gap by providing continuity for NOAA's observational requirements in the afternoon polar orbit. These include:

- **S-NPP mission life extension:** This involves gathering lessons learned from long-lived NOAA and Goddard Space Flight Center missions that can be applied to S-NPP to prevent a near-term gap from occurring. It includes reviewing satellite monitoring and trending processes, identifying life-limiting items and consumables, assessing the status of residual risk and watch items, and ensuring the robustness of operations in response to on-orbit component failures.
- **Timely and successful JPSS-1 launch:** This involves taking the necessary steps to ensure the timely and successful launch of the JPSS-1 satellite. This includes keeping JPSS-1 within budget as well as continuing to aggressively manage the program's progress towards its launch date.

- **JPSS-2 launch readiness acceleration:** This involves accelerating the development of the JPSS-2 satellite to reduce the likelihood of a potential gap should there be issues with the JPSS-1 satellite. It includes evaluating proposals and negotiating instrument contracts.
- **"Gap filler" mission:** Recommended by an independent review team, this alternative involves the potential implementation of "gap filler" satellites that would carry two key environmental instruments: ATMS and CrIS. The purpose of the satellites would be to bridge a gap in polar satellite data due to S-NPP failure, JPSS-1 launch failure, or JPSS-1 spacecraft failure.
- **SIDAR mission:** This alternative is intended to bridge the gap between satellites currently providing search and rescue, data collection, and solar irradiance capabilities in the afternoon polar orbit. It provides observations that were once part of the JPSS mission.

These alternatives have multiple benefits and challenges. Benefits include providing continuity for the S-NPP and JPSS-like observations. Challenges include the time it would take to develop the gap-filler instruments as well as the funding necessary to acquire them. Also, NOAA has not yet decided whether and how it will establish the SIDAR mission. Table 5 describes the key benefits and challenges commonly identified for alternatives that could reduce the likelihood of a potential afternoon polar data gap.

Table 5. Key Benefits and Challenges for Alternatives that Could Reduce the Likelihood of an Afternoon Polar Data Gap

Alternatives	Key benefits	Key challenges
S-NPP mission life extension	• Provides continuity of JPSS-like observations	• Satellite may not last as long as expected
	• Minimizes impact to users and operations	
Timely and successful JPSS-1 launch	• Provides continuity of JPSS-like observations • Minimizes impact to users and operations	• JPSS-1 may not be technically and mechanically ready for launch
JPSS-2 launch readiness acceleration	• Provides continuity of JPSS-like observations	• The spacecraft and instruments may not be available in time for an early launch

Table 5. (Continued)

Alternatives	Key benefits	Key challenges
	• Reduces the gap if there was a JPSS-1 failure	• This option may require additional funding
"Gap filler" mission recommended by the National Environmental, Satellite, Data, and Information Service (NESDIS) Independent Review Team	• Provides continuity of JPSS-like observations (ATMS/CrIS) • Reduces the gap if there was a JPSS-1 failure	• The spacecraft and instruments may not be available in time for an early launch •This option would require funding
NOAA's SIDAR mission	• Provides continuity of polar observations (TSIS, A-DCS, SARSAT)	• This option would require funding

Source: GAO analysis based on interviews with experts from weather satellite-related organizations. | GAO-15-47

Note: A-DCS—Advanced Data Collection System; ATMS—Advanced Technology Microwave Sounder; CrIS—Cross-Track Infrared Sounder; JPSS—Joint Polar Satellite System; JPSS-1—Joint Polar Satellite System-1; JPSS-2—Joint Polar Satellite System-2; NOAA—National Oceanic and Atmospheric Administration; SARSAT—Search and Rescue Satellite-Aided Tracking system; SIDAR—Solar Irradiance, Data, and Rescue; S-NPP—Suomi National Polar-orbiting Partnership; and TSIS—Total and Spectral Solar Irradiance Sensor.

Alternatives for Reducing the Impact of Satellite Data Gaps in the Afternoon Polar Orbit Offer Unique Benefits and Challenges

The alternatives that could reduce the impact of a potential afternoon polar data gap include both space-based and non-space-based mitigation options. This category of alternatives can be separated into the following subcategories: (a) extend and expand the use of current data sources; (b) enhance modeling and data assimilation; (c) develop new data sources; and (d) explore opportunities with foreign and domestic partners. The following provides more details on each of the subcategories.

- **Extend and expand the use of current data sources:** This includes alternatives for observations that are currently available and in use and that, if extended or expanded, might mitigate the impact of an afternoon polar data gap. For instance, extending the life of legacy satellites—such as POES and Aqua—could provide similar, but less capable, backup observations if there were a gap. Also, extending and expanding other current data sources that are already top contributors to numerical forecast models, such as aircraft observations from domestic and international commercial flights and Global Positioning System radio occultation measurements, could mitigate the impact of a gap by providing more observations. Additionally, continuing to rely on and improve NOAA's use of observations from other polar satellites outside the afternoon orbit, such as DOD's DMSP and European Metop satellites, could also help alleviate the impact of a gap in the afternoon polar orbit by providing similar environmental observations that occur at different times of the day.

- **Enhance modeling and data assimilation:** This includes alternatives that enhance the extraction of information from existing observed data, including data assimilation and modeling, increased operational computing capability, and multi-model ensemble products. Enhancing this back-end data processing could help mitigate the impact of a polar-orbiting satellite data loss in the afternoon orbit by making numerical forecast models more efficient and effective in processing data that continues to be received. For example, employing 4-dimensional data assimilation could improve the use of observations by using better quality assurance and higher resolution analysis. Also, accelerating the next-generation global forecast model from research into operations would mitigate forecasting skill loss by introducing more advanced modeling techniques and higher resolution. Additionally, blending global models could mitigate the impact of an afternoon gap in observations by adjusting the weights of the other global models used to influence the forecast to minimize the degradation in NOAA weather services. Finally, adding additional direct readout sites could make the remaining observations more useful by reducing data latency (the time it takes for data to reach the forecast models) potentially allowing for more observations to be used in each model run.
- **Develop new data sources:** This includes alternatives that are hypothesized to have a positive impact on weather forecast models but need more time for development and testing before they are technically and operationally ready. For instance, data from

geostationary satellite imagery-derived moisture profiles, atmospheric motion vectors (imagery-derived wind information), cloud-impacted infrared and microwave soundings, and doppler wind lidar could contribute to better forecast model performance and possibly reduce model dependence on existing data sources. Also, the targeted observations from manned and unmanned aircraft could provide a proving ground for performing demonstrations with new instruments that may mitigate the loss of polar satellite observations.

- **Explore partnership opportunities:** These alternatives focus on the partnership between NOAA and another party (i.e., commercial vendors, foreign government, NASA, DOD, and academia) to help mitigate a potential gap in the afternoon polar orbit. This includes alternatives such as accelerating the procurement of sensors, satellites, and launch vehicles; contracting for small satellites provided by industry; commercial data buys; hosted payloads; and prototype-stage technologies. Additionally, creating partnership agreements with foreign governments to potentially acquire data from satellites like China's Feng-Yun 3 series, Canada's proposed Polar Communications and Weather mission, and European Sentinel satellites could help reduce the impact of a potential gap by acquiring data that NOAA could use in weather forecast models, obtaining imagery in higher latitude and polar regions, or collecting data for processing environmental products.

There are key benefits and challenges associated with these alternatives. A key benefit is that some observations are already in use. For instance, aircraft observations and Global Positioning System radio occultation are already top contributors to numerical weather prediction models' error reduction, and by extending and expanding their use, the models should become more resilient in the event of a gap. Also, the alternatives that enhance modeling and data assimilation could also make NOAA's weather models more effective in dealing with a data gap. Key challenges involve timeliness. Because new data sources still need to be developed and tested, they may not be operational in time for a potential near-term gap. In addition, alternatives that require partnership agreements may not address a potential near-term gap due to data processing and data sharing concerns. Table 6 describes the key benefits and challenges commonly identified for alternatives that could reduce the impact of a potential gap.

Table 6. Key Benefits and Challenges for Alternatives that Could Reduce the Impact of an Afternoon Polar Data Gap

Subcategory	Key benefits	Key challenges
Extend and expand use of current data sources Examples: extension of legacy satellites (POES, Aqua), DMSP, midmorning orbit data from European Metop satellites, aircraft observations from domestic and international commercial flights, and Global Positioning System radio occultation data.	• Legacy satellites' observations, even though of lesser quality, could help partially mitigate the impact of potential gaps • Users are currently using and familiar with data sources • Global Positioning System radio occultation is an "anchor measurement" that improves the accuracy of other data sources in numerical weather prediction models • Aircraft observations and Global Positioning System radio occultation are top contributors to numerical weather prediction models' error reduction after polar microwave and infrared sounders	• Legacy polar sounders and imagers are not the same quality as ATMS, CrIS, and VIIRS • Legacy satellites may not be available when a gap occurs • Aircraft observations are limited by flight paths, which limit global and vertical coverage • Certain alternatives (i.e. DMSP, Metop) are not in the afternoon polar orbits
Enhance modeling and data assimilation Examples: 4-dimensional data assimilation, accelerating global model research to operations, increasing high-performance computing capacity for research and operations, blending of global models, and adding additional direct readout sites to reduce data latency.	• 4-dimensional data assimilation makes weather forecast models more robust with fewer observations • The next-generation global forecast model could improve NOAA's forecast skill with or without a gap • Other alternatives' development are enabled by additional research and operational high-performance computing	• Alternatives do not replace the need for observations • High-performance computing capacity for research and operations is in high demand • 4-D data assimilation and accelerating the next-generation global forecast model may not be operational in time for a potential gap before JPSS-1 launches

Table 6. (Continued)

Subcategory	Key benefits	Key challenges
Develop new sources Examples: geostationary satellite imagery-derived moisture profiles, atmospheric motion vectors (imagery-derived wind information), cloud-impacted infrared and microwave soundings, doppler wind lidar, and targeted observations from unmanned aircraft.	• New data sources are promising in terms of mitigating the impact of data gaps by providing additional observations to the models • Non-NOAA weather organizations are already using certain data sources operationally	• New data sources can take one year or longer to transition to operations after the science is developed • Certain data sources' mitigation value may not prove to be worth implementing operationally • Unmanned aircraft may have difficulty targeting weather events as they develop (which is key for weather forecast model contribution)
Explore partnership opportunities Examples: accelerated procurement (sensors, satellites, and launch vehicles), contracting for small satellites provided by industry, commercial data buys, hosted payloads, prototype-stage technologies, China's Feng-Yun 3 series, Canada's proposed Polar Communications and Weather mission, European Sentinel satellites, and various foreign satellites with search and rescue sensors.	• • Foreign satellites could provide more observations into models • Foreign satellites may offer instruments and capabilities that are comparable to JPSS program requirements • Commercially developed solutions may provide future or long-term options	• New foreign partnership agreements may not address a near-term gap • NOAA requires agreements related to open data sharing Partnerships increase potential data security risks • New foreign partnership agreements increase reliance on foreign governments for observations • Commercial data buys and hosted payloads have limited options in the short-term • Foreign SARSAT and A-DCS capabilities do not cover the polar afternoon orbit

Source: GAO analysis based on interviews with experts from weather satellite-related organizations. | GAO-15-47

Note: A-DCS—Advanced Data Collection System; ATMS—Advanced Technology Microwave Sounder; CrIS—Cross-Track Infrared Sounder; DMSP—Defense Meteorological Satellite Program; JPSS—Joint Polar Satellite System; JPSS-1—Joint Polar Satellite System-1; Metop—Meteorological Operational (satellite); NOAA—National Oceanic and Atmospheric Administration; SARSAT—Search and Rescue Satellite-Aided Tracking system; and VIIRS—Visible Infrared Imaging Radiometer Suite.

Key Alternatives May Not be Available in Time to Address a Near-term Data Gap

While there are multiple alternatives available for preventing and reducing the impact of a gap, most of these options may not be available soon enough to address a near-term gap. In terms of preventing the potential gap, NOAA is currently doing what it can to extend the S-NPP mission life and to ensure the timely and successful JPSS-1 launch. However, the other alternatives for preventing a gap (including accelerating JPSS-2 launch readiness and acquiring a "gap filler" satellite) are not viable in the near-term because these instruments take years to develop and test. Additionally, NOAA is not currently funding the SIDAR mission.

Alternatives that extend and expand the use of current data sources are viable solutions to reduce the impact of a potential gap because users are familiar with and using the existing data sources. However, alternatives for enhancing modeling and data assimilation, developing new sources, and exploring partnership opportunities face challenges that may restrict them from reducing the impact of a potential near-term gap. For instance, it would take additional high-performance computing capacity to enhance modeling and data assimilation. Also, developing new data sources may take additional time to transition to operational use after the research is finished. In addition, NOAA would need to establish agreements about data sharing, data processing, and information security before embarking on new foreign partnership agreements, which may be subject to legal restrictions. Finally, while long-term commercial partnerships could address satellite data needs, these options will not likely be available in time to address a near-term gap.

While all of the alternatives have trade-offs, several alternatives may represent the best known options for reducing the impact of a gap:

- Extending legacy satellites like POES and Aqua, continuing to obtain data from European midmorning satellites, and ensuring legacy and European satellites' data quality remains acceptable would be a strong fallback strategy because forecasters and their models have been using these data operationally for years. Moreover, these alternatives could be important in the near-term when other options may not be available.
- Obtaining additional observations of radio occultation and commercial aircraft data are likely to provide more significant mitigation than other data sources because studies conducted by leading forecast modeling centers have validated that they are among the top contributors for reducing forecast errors.
- Advancing 4-dimensional data assimilation and the next generation global forecast model are likely to provide significant mitigation because they are expected to make more efficient use of data that are still available and produce improved techniques for evaluating data with each model run.
- Increasing high-performance computing capacity is likely to provide significant mitigation because it is the key factor for enabling greater resolution in existing and future models and it drives the pace of development for assimilation of new or additional data that could further improve NOAA's models.

As discussed in the next section, NOAA is currently pursuing all of these alternatives, although it has had mixed progress in implementing them.

NOAA Improved its Contingency Plan, but Delays May Limit the Effectiveness of Key Mitigation Activities and Oversight is Inconsistent

Government and industry best practices call for the development of contingency plans to maintain an organization's essential functions in the case of an adverse event.[24] As a complement to risk mitigation, contingency planning includes strategies that attempt to reduce or control the impact of risks should they occur. These practices identified by, for example, the National Institute of Standards and Technology and the Software Engineering Institute, include key elements such as defining failure scenarios, identifying and selecting strategies to address failure scenarios, developing procedures and actions to implement the selected strategies, testing the plans, and involving affected stakeholders. These elements can be grouped into categories, including (1) identifying failure scenarios and impacts, (2) developing contingency plans, and (3) validating and implementing contingency plans (see table 7).

Table 7. Guidelines for Developing a Sound Contingency Plan

Category	Description of key activities
Identifying failure scenarios and impacts	Includes defining failure scenarios; conducting impact analyses that show the impact of failure scenarios; defining minimum acceptable levels of outputs and recovery time objectives; and establishing resumption priorities.
Developing contingency plans	Includes identifying alternative solutions to address failure scenarios; selecting contingency strategies from among alternatives based on costs, benefits, and impacts; defining actions, roles and responsibilities, triggers, and timelines for implementing contingency plans; developing "zero-day" procedures; ensuring that steps reflect priorities for resumption of products and recovery objectives; and obtaining review and approval of the contingency plan from designated officials.
Validating and implementing contingency plans	Includes identifying steps for testing contingency plans and conducting training exercises; preparing for and executing tests; validating test results for consistency against minimum performance levels; executing applicable actions for implementation of contingency strategies; communicating and coordinating with stakeholders to ensure that the strategies remain optimal for reducing potential impacts; and updating and maintaining contingency plans as warranted.

Source: GAO analysis of guidance documents from the National Institute of Standards and Technology, Software Engineering Institute, and GAO. | GAO-15-47

NOAA developed its original polar satellite gap contingency plan in October 2012. We reported in September 2013 that NOAA had not yet selected the strategies from its plan to be implemented or developed procedures and actions to implement the selected strategies.[25] Further, NOAA's initial plan had shortfalls when compared to government and industry best practices. NOAA agreed with our recommendation to establish a comprehensive contingency plan for potential satellite data gaps in the polar orbit that is consistent with contingency planning best practices.

In February 2014, NOAA updated its polar satellite gap contingency plan. NOAA's revised plan focuses on gap mitigation for weather services through activities in three primary categories: (1) understanding the probability and impacts of a gap, (2) preventing potential gaps, and (3) reducing the impact of potential gaps. The plan also includes actions for maintaining other important capabilities in the polar afternoon orbit, including continuing the total solar irradiance data record, continuing data collection, and continuing search and rescue support.

NOAA has made several notable improvements with its recent plan update. For example, NOAA has expanded its plan to include additional alternatives that experts identified. In addition, the plan accounts for additional gap scenarios such as the loss of early morning DMSP satellite data and midmorning European satellite data; integrates other alternatives that were identified from a prior study; identifies the mitigation strategies to be executed; and identifies specific activities with roles and responsibilities, triggers, and deadlines for implementing those strategies.

However, additional work remains for NOAA's contingency plan to fully address government and industry best practices for contingency planning. For example, NOAA's plan does not yet include an assessment of the available alternatives based on their cost and potential impacts and testing remains to be executed in order to validate that the agency has the optimal contingency plan. The following table describes previously identified weaknesses in NOAA's October 2012 contingency plan and the extent to which the agency has made progress in addressing the shortfalls in its latest plan.

Table 8. NOAA's Progress in Polar Satellite Contingency Planning

October 2012 contingency planweaknesses	Description of progress in February 2014 plan
Category: Identifying failure scenarios and impacts	
The plan did not address certain scenarios such as the possibility of a loss of data from DOD and European partner satellites in morning orbits or a Japanese partner mission in the afternoon orbit.	Progress: Partially addressed The revised plan states that NOAA reviewed its analyses and available information and determined that there were no other options to DOD and European satellites in the morning orbits and the options in its current contingency plan will serve as mitigation against the potential loss of this data. NOAA also stated that it participated in the Air Force Space Command analysis of alternatives so that it is cognizant of DOD follow-on satellite plans and how they may affect NOAA's plans.

October 2012 contingency planweaknesses	Description of progress in February 2014 plan
	However, the Air Force has not finalized its plans for a weather follow-on mission. Thus, NOAA does not yet have full knowledge about the probability of data availability or capabilities to be provided by future DOD polar satellites. The plan notes that the agency has had a memorandum of agreement in place with Japan since 2011 to obtain microwave imagery from its satellites. According to NOAA, the current Japanese satellite is expected to last until at least 2017 and two more missions are to provide continuity for this data through 2025. However, NOAA has also identified risks to Japan's ability to secure funding for these next two satellites with no comparable sensor to support 13 JPSS product requirements.
The plan did not include recovery time objectives for key data products.	Progress: Not addressed The plan states that future updates will provide information on recovery time objectives for key data products.
Category: Developing contingency plans	
The plan did not identify the contingency strategies NOAA selected to be implemented or establish procedures and actions to implement the selected strategies.	Progress: Partially addressed NOAA's revised plan identifies approximately 30 contingency strategies to be implemented in four categories: (1) understanding the probability and impacts of a gap; (2) preventing potential gaps; (3) reducing the impact of potential gaps; and (4) maintaining other capabilities. The plan includes procedures and actions to implement the selected strategies in each of these areas, such as conducting observing system experiments to replicate conditions during a gap period; installing additional high-performance computing capacity; extending mission life for the current polar satellite; and assimilating new observations on clouds, commercial aircraft, and other data.a However, the plan appears to be incomplete. NOAA has three ongoing gap mitigation projects that are not included in the polar satellite gap contingency plan. They are (1) S-NPP data processing and distribution; (2) the Hurricane Forecast Improvement Project; and (3) moving the Hurricane Forecast Improvement Project from research to operations. In addition, there are other alternatives identified by experts that are not included in NOAA's contingency plan.

Table 8. (Continued)

October 2012 contingency planweaknesses	Description of progress in February 2014 plan
	These include: (1) developing doppler wind lidar, (2) developing products from NASA's Global Precipitation Measurement mission, and (3) exploring the use of the International Space Station as a host for instruments.
The plan had not yet been integrated with other alternatives that were subsequently identified.	Progress: Fully addressed The revised plan integrates alternatives identified in NOAA's October 2012 plan with additional options identified in a subsequent study.
NOAA had not yet assessed its alternative strategies based on costs, benefits, and potential impacts.	Progress: Not addressed The plan does not provide details on NOAA's assessment of the available alternatives based on their costs, benefits, and potential impacts.
The plan did not identify options for preventing gaps from occurring.	Progress: Fully addressed The plan currently identifies mitigation steps being taken or to be taken to prevent gaps from occurring.
The plan did not identify opportunities for accelerating the calibration and validation phase—the time between launch and availability of operational products—for JPSS-1.	Progress: Partially addressed We reported in September 2013 that NOAA had experienced delays in validating key products for S-NPP. NOAA's revised plan states that the calibration and validation phase for JPSS-1 key performance parameter data (ATMS, CrIS, and VIIRS) will be 6 months. Program officials stated that they believe calibration times will be less for JPSS-1 due to the collection of S-NPP on-orbit data and their experience in calibrating and validating S-NPP instruments and data, including lessons learned. However, it has taken the program over 2 years to calibrate and validate S-NPP data and we have concerns as to whether calibration/validation can be completed in 6 months. According to NOAA, future plan updates are to provide an assessment of opportunities for further accelerating the calibration and validation phase on JPSS-1 and JPSS-2.
The plan did not always identify specific actions with defined roles and responsibilities, timelines, and triggers.	Progress: Partially addressed The plan identifies specific actions for mitigation strategies, general roles and responsibilities, line offices that are responsible for implementing specific actions, and triggers that are to result in actions or decisions when certain events occur.

October 2012 contingency planweaknesses	Description of progress in February 2014 plan
	Among other scheduling best practices, an exposure draft of GAO's Schedule Assessment Guide states that durations should be reasonably short and meaningful and allow for discrete progress measurement.b However, the timelines in NOAA's plan do not provide for meaningful progress measurement because activity deadlines are long in duration—by fiscal year or longer—while other activities have no deadlines. GAO's Schedule Assessment Guide also states that activities that must be completed before other activities can begin (predecessor activities), as well as activities that cannot begin until other activities are completed (successor activities), should be identified. NOAA's plan does not clearly identify activities that need to be completed before other activities can begin, such as high performance computing upgrades that are required for observing system experiments and assimilation of new data sources. There is no integrated master schedule that identifies such linkages across NOAA's polar gap mitigation activities. Because deadlines are long in duration or do not exist for certain projects and there is no schedule integrating mitigation activities that are dependent upon each other, NOAA's ability to measure progress and ensure its planned activities remain on schedule is constrained.
Category: Validating and implementing contingency plans	
NOAA had not yet initiated efforts to validate or implement its gap mitigation plan.	Progress: Partially addressed NOAA has begun implementing its plan. Initial steps to acquire data, develop the science for new data, install technology and infrastructure upgrades, and hire subject matter experts are generally planned for fiscal year 2014 while validation activities are generally to be performed in fiscal year 2015 or later. However, completion dates for testing and validating the alternatives are not always defined. Until additional testing and validation activities are executed, NOAA will not have full knowledge regarding the potential impact of its planned contingency actions.

Source: GAO analysis of NOAA data. | GAO-15-47

[a]Observing system experiments provide objective assessments of existing observing systems' contributions to numerical weather forecasts while observing system simulation experiments provide objective assessments of the contributions to numerical weather forecasts from future potential observing systems.

[b] *GAO Schedule Assessment Guide: Best Practices for Project Schedules*, GAO-12-120G (exposure draft) (Washington, D.C.: May 30, 2012).

NOAA's contingency plan and NOAA officials acknowledge several of these shortfalls, including the need to identify recovery time objectives, identify opportunities for further accelerating the calibration and validation phase for JPSS-1 and JPSS-2, update trigger points, and review options for testing and validating the plan. The plan and agency officials stated that these elements would be added to future updates that are to occur on a semiannual cycle. In addition, the NESDIS official responsible for updating and reporting on the status of NOAA's contingency plan stated that the plan did not include details on NOAA's assessment of the available alternatives based on their costs, benefits, and potential impacts because decisions on those alternatives were still being deliberated at the time of the last plan update. Also, a National Weather Service scientist who played a key role in selecting NOAA's gap mitigation strategies stated that, while the agency can make educated guesses about the potential impacts to numerical forecast systems for certain alternatives, scientists will not be able to validate the impact of these mitigation strategies until the strategies are ready for the implementation phase and are tested for their individual contributions in an operational forecast system. Based on current schedules, such testing is generally not expected to occur until fiscal year 2015 or later for National Weather Service gap mitigation strategies. Until NOAA fully addresses the elements identified above to improve its contingency plan, it may not be sufficiently prepared to mitigate potential gaps in polar satellite coverage.

NOAA Has Begun Implementing its Contingency Plan, but Key Activities Have Encountered Setbacks and Efforts to Monitor Progress Are Not Consistent or Comprehensive

While NOAA has started to implement its polar satellite gap contingency plan, it has experienced challenges in implementing key activities outlined in the plan. From the variety of alternatives noted earlier, NOAA has identified 21 mitigation projects that are to be implemented in order to address the potential for satellite data gaps in the afternoon polar orbit. These projects are supported primarily from the weather satellite data gap mitigation reserve fund from the Disaster Relief Appropriations Act, 2013. Table 9 summarizes the gap mitigation projects NOAA is implementing.

Table 9. Polar Gap Mitigation Projects Being Implemented by NOAA

Project	Budget ($M)	Original estimated completion	Key expected outcomes
Operational high-performance computing to accommodate data assimilation upgrades and observation enhancements	$18.6M	Dec 2014	• Increase horizontal resolution of the Global Forecast System (NOAA's primary forecast model) from 27 to 10 kilometers through 10-day forecasts • Make similar improvements to other models
POES/Metop data continuity	$17.4M	Aug 2015	• Ensure continuity of the key performance parameter measurements from Metop
Augmenting research high-performance computing resources and software engineering	$15.0M	Apr 2015	• Rebalance research and development capacity to approximately twice that of operations to support experimental runs of higher resolution, more complex models • Provide support for other gap mitigation activities
Hurricane Forecast Improvement Project	$10.1M	Mar 2015	•Improve numerical hurricane forecast system to reduce errors in track and intensity by 50% (20% in first 5 years) •Extend forecast skill beyond 7 days •Improve prediction of rapid intensification, reduce false alarms •Reduce overall preventable losses
Targeted observations for high impact events	$9.0M	Dec 2016	• Quantify the significance of unmanned observations to high impact weather prediction • Quantify the cost and operational benefit of unmanned observing technology for high impact weather prediction

Table 9. (Continued)

Project	Budget ($M)	Original estimated completion	Key expected outcomes
Aircraft observations	$9.0M	Jan 2016	• Increase accuracy of numerical weather prediction for high impact events • Quantify the value of aircraft observations • Increase forecast skill performance by 1.0%
Hurricane Forecast Improvement Project: research to operations	$3.6M	Mar 2015	Same as Hurricane Forecast Improvement Project
Observing system simulation experiments	$3.3M	Dec 2015	• Develop the next generation global observing system simulation experiment system to replace the current obsolete system • Conduct preliminary observing system experiments and observing system simulation experiments in support of data gap mitigation
S-NPP data processing and distribution	$3.0M	Sep 2015	Enable 24/7 data processing for S-NPP (currently 8 hours/5 days per week)
Geostationary data (GOES-R and international missions)	$2.7M	Nov 2015	• Provide new operational cloudiness product for users • Increase forecast skill performance by 0.25%
Assimilation of cloud-impacted radiances, quality control, observation error, and surface emissivity formulations	$2.5M	Nov 2015	• Increase the amount and spatial coverage of data assimilated for operations • Reduce the probability of a poor forecast • Increase forecast skill performance by 2.0%

Project	Budget ($M)	Original estimated completion	Key expected outcomes
Blends of global models	$2.0M	Dec 2015	• Provide improved verification product set for future use • Increase forecast skill by 1.0%
High-performance computing hardware augmentation for observing system simulation experiments	$2.0M	Dec 2015	• Accelerate research to operations activities with gap mitigation value • Support data denial experiments to assess numerical weather prediction impacts from potential data gaps • Support assessments of potential impact of future sensors on numerical weather prediction
Dual X/L band direct broadcast receipt sites	$1.7M	July 2016	• Install three new direct broadcast antennas at Monterey, California; Miami, Florida; and Puerto Rico • Reduce latency of sounder data from polar-orbiting satellites to less than 15 minutes • Provide direct broadcast users with satellite sounder products
Direct readout imagery from other satellites	$1.7M	Aug 2015	• Transform direct broadcast research system in Alaska to operational capability • Ensure reliable flow of satellite data to Alaska operations
Atmospheric motion vectors	$1.0M	Nov 2015	•Transfer atmospheric motion vector algorithms developed by the Navy and University of Wisconsin into NOAA operations • Increase forecast skill by 0.5%

Table 9. (Continued)

Project	Budget ($M)	Original estimated completion	Key expected outcomes
High-performance computing software management and integration	$0.9M	Dec 2014	Same as operational high-performance computing project
Geostationary data (GOES-R and international)	$0.8M	Nov 2015	• Ensure Himawari-8 advanced imagery data are available to numerical weather prediction models
4-dimensional data assimilation	$0.8M	Nov 2015	• Improve use of observations by using better quality assurance and higher resolution • Increase computational efficiency and reduce code components • Increase forecast skill by 1.5%
Observing system experiments	$0.2M	Nov 2015	• Run regional experiments to replicate model conditions with no polar afternoon satellite data • Identify quantitative impact of data gap and latency on standard forecast scores
DMSP Special Sensor Microwave Imager/Sounder	$0.1M	Sep 2014	• Assimilate DMSP Special Sensor Microwave Imager/Sounder data into NOAA systems

Source: GAO analysis based on NOAA data. | GAO-15-47

Note: DMSP—Defense Meteorological Satellite Program, GOES-R—Geostationary Operational Environmental Satellite-R series, Metop—Meteorological Operational (satellite), NOAA—National Oceanic and Atmospheric Administration, POES—Polar-orbiting Operational Environmental Satellites, and S-NPP—Suomi National Polar-orbiting Partnership.

NOAA has demonstrated progress by implementing initial activities on these gap mitigation projects. The agency offices responsible for these projects have reported completing initial milestones such as defining statements of objectives, developing spending plans, obtaining funding, and getting the necessary workforce in place for execution. In addition, certain mitigation projects, such as the Hurricane Forecast Improvement Project and observing system experiments, have already begun. NOAA also completed a key upgrade to two supercomputers to be used for observing system simulation experiments.

However, NOAA has experienced delays in executing other key activities. For example:

- A planned upgrade to the National Weather Service's (NWS) operational high-performance computing was to occur by December 2014, but officials stated that the planned upgrade could be delayed by 12 months or longer to implement risk mitigation activities associated with the sale of a vendor's server business. This concern is expected to not only affect this individual project but has the potential to affect other NWS projects that were counting on having these additional computing resources in place in order to test and validate their actual impact on numerical weather prediction models prior to being implemented operationally. NWS officials stated that they are planning to partially address this issue by implementing an interim upgrade for operational high-performance computing in February 2015. The interim upgrade is expected to provide about half of the total planned computing capacity for planned enhancements to NWS weather model resolution and for testing and implementation of new data sources that are mature enough in their development. Officials reported that other data sources that require further development and further enhancements of model resolution will be supported by the full upgrade that is to be completed by July 2016.
- NOAA does not plan to complete observing system experiments that are to provide quantitative information regarding the anticipated degradation to NOAA's numerical weather prediction models from the absence of afternoon polar-orbiting satellite data until March 2016, which is 4 months later than planned. A contractor NOAA engaged in October 2012 to study alternatives for polar afternoon gap mitigation stressed the importance of conducting these experiments as soon as possible in order to facilitate informed decision making in choosing the best mitigation strategy. NOAA officials stated that a global observing system experiment has been completed that provides estimated quantitative impact from JPSS data, but regional

experiments remain to be completed. Thus, NOAA has been planning and may continue to implement gap mitigation measures over the next 2 years with limited information about the potential problem it is attempting to address. According to NWS progress reports, this project has been delayed in part because the computing resources the project received were 60 percent less than required.

- Multiple projects have been affected by a major shortfall in the availability of high-performance computing for research and development efforts during fiscal year 2014. According to a report by NOAA's high-performance computing working group, users' original request exceeded available resources by 256 percent due primarily to a spike in computing required to execute new and expanded gap mitigation efforts. The working group reported that it found a way to provide some allocation to all requesting parties but none of those parties received the allocation needed to meet their 2014 mission objectives.

Because a potential near-term data gap could occur sooner and last longer than expected, NOAA's ongoing gap mitigation efforts are becoming even more critical. According to Office of Management Budget guidance, projects that require extensive development work before they can be put into operation are inherently risky and should be prioritized by comparing their costs and outcomes to other projects within a portfolio.[26] Further, our previous work has found that projects should be prioritized when an agency is at risk of not meeting key deadlines and several strategies exist to accelerate schedules.[27]

However, the agency has not prioritized or accelerated activities most likely to address a gap because they have been focused on moving out on many different initiatives to see which ones will have the most impact. NOAA officials stated that further prioritization among mitigation activities was not warranted because the activities were fully funded and were not dependent on the completion of other activities. We disagree. There are dependencies among projects that would benefit from prioritization. For example, developing and utilizing new data sources is dependent on improving the capacity of high-performance computing. While it makes sense to investigate multiple mitigation options, unless NOAA assesses the activities that have the most promise and accelerates those activities, it may not be sufficiently prepared to mitigate near-term data gaps.

Efforts to Monitor Progress of Contingency Planning Activities Are Not Consistent or Comprehensive

Three NOAA entities—NESDIS, NWS, and the Office of Oceanic and Atmospheric Research (OAR)—have primary responsibilities for implementing NOAA's February 2014 polar satellite gap contingency plan and associated mitigation activities. In order to provide oversight and monitor the budget, milestones, and performance of the agency's gap mitigation efforts, NOAA's contingency plan requires the officials responsible for contingency activities to provide monthly progress briefings to the assistant administrators for each of the three NOAA entities. Further, the plan requires them to provide quarterly progress briefings to NOAA's program management council.

While NOAA is providing some oversight of its many gap mitigation projects and activities, the agency's oversight efforts are not consistent or comprehensive. Specifically, NWS is providing the required monthly reporting on the status of its gap mitigation projects. However, NESDIS and OAR are not providing these reports. NOAA officials stated that monthly progress reporting was something that NWS did on its own and that in lieu of monthly reporting, OAR and NESDIS would provide quarterly reports on their respective mitigation projects. However, this is inconsistent with the requirement in NOAA's contingency plan. Until NOAA provides monthly reporting for all gap mitigation projects, the agency's oversight of these projects will be constrained.

In addition, responsible officials are providing quarterly progress briefings on 21 mitigation projects to NOAA's program management council. However, the responsible entities have not reported on 9 mitigation activities outlined in NOAA's contingency plan because they were not identified among the 21 mitigation projects. Examples of the 9 projects include expanding Global Positioning System radio occultation measurements; and projects related to continuing ozone, solar, search and rescue, and data collection capabilities. NOAA officials explained that selected projects are not being tracked the same as the 21 mitigation projects because they are funded under different appropriations. Because of the criticality of efforts to mitigate the potential gap and the need for oversight of these efforts, it is important for NOAA to monitor the progress being made on activities no matter their funding source.

Since these projects are not included in the monthly and quarterly progress reports that are required by NOAA's contingency plan, the extent to which they are being implemented and the extent to which they could effectively mitigate potential data gaps in the polar afternoon orbit may not be effectively monitored. Until NOAA improves the consistency and comprehensiveness of its progress monitoring, the agency's ability to ensure that the necessary progress is being made will be limited.

CONCLUSION

Facing an unprecedented gap in weather satellite data—which could have a devastating effect on the nation's economy and the safety of its population—NOAA is working to launch its next polar-orbiting environmental satellite as soon as possible. However, the JPSS program is facing risks and it has not yet tracked actual dates for completing activities in mitigating those risks. Further, the program revised its estimate of how long a gap could last down to 3 months, but this estimate was based on inconsistent and unproven assumptions and did not account for the risk that space debris pose to S-NPP's life expectancy. Until it does so, the program's ability to monitor progress in reducing significant risks and to understand the potential for a near-term gap will be limited.

Experts have identified multiple alternative ways to help mitigate the potential gap. These alternatives include taking steps to prevent and limit the gap as well as steps to reduce the impact of a gap once it occurs. While the alternatives promise increased robustness of the polar satellite infrastructure, key alternatives are not likely to be available in time should a near-term gap occur.

Taking these many alternatives into consideration, NOAA has improved its contingency plan and is moving forward to implement multiple mitigation projects. However, it has not yet addressed shortfalls such as providing key information about alternatives' costs and impacts, and establishing when testing of selected alternatives is to be completed. In addition, while the agency is moving forward on numerous initiatives, it has not prioritized them. NOAA has years of work remaining in order to fully implement and validate the data that the alternative sources offer, and it has already experienced persistent setbacks in completing mitigation efforts because of issues in obtaining sufficient computing resources. These factors make it more unlikely that NOAA will complete key mitigation activities before a gap occurs. Until NOAA prioritizes and accelerates mitigation activities with the greatest potential to reduce the impact of gap to weather forecasting, it may not be sufficiently prepared to mitigate near-term data gaps. Moreover, while NOAA is providing some oversight of its many gap mitigation projects and activities, the agency's oversight efforts are not consistent or comprehensive. Until NOAA demonstrates that it is making swift and effective progress in mitigating potential near-term gaps in polar satellite data and that it is effectively overseeing these important efforts, there will be a growing risk that degraded forecasts and warnings will lead to negative impacts on the health and safety of the U.S. population and economy.

RECOMMENDATIONS

Given the importance of reducing key risks to JPSS satellites we are making the following 5 recommendations to the Secretary of Commerce. Specifically, we recommend that the Secretary direct the Administrator of NOAA to

- track completion dates for all risk mitigation activities; and
- update the program's assessment of potential polar satellite data gaps to include more accurate assumptions about launch dates and the length of the data calibration period, as well as key risks such as the potential effect of space debris on JPSS and other polar satellites' expected lifetimes.

In addition, because NOAA has not fully implemented our prior recommendation to establish a comprehensive contingency plan for potential satellite data gaps in the polar orbit that is consistent with contingency planning best practices, we recommend that the Secretary direct the Administrator of NOAA to

- revise the polar satellite contingency plan to address the shortfalls noted in this report, such as identifying DOD's and Japan's plans to continue weather satellite observations, including recovery time objectives for key products, completing the contingency plan with selected strategies, identifying opportunities for accelerating calibration and validation of products, providing an assessment of available alternatives based on their costs and potential impacts, establishing a schedule with meaningful timelines and linkages among mitigation activities, and defining completion dates for testing and validating the alternatives.

Additionally, in order to enhance NOAA's preparedness for potential polar satellite data gaps in the near-term, we recommend that the Secretary direct the NOAA Administrator to direct the Assistant Administrators of NESDIS, NWS, and OAR to

- investigate ways to prioritize mitigation projects with the greatest potential benefit to weather forecasting in the event of a gap in JPSS satellite data and report recommendations to the NOAA program management council; and
- ensure that the relevant entities provide monthly and quarterly updates on the progress on all mitigation projects and activities during existing monthly and quarterly management meetings.

AGENCY COMMENTS AND OUR EVALUATION

We sought comments on a draft of our report from the Department of Commerce, the Department of Defense, and NASA. We received written comments from the Deputy Secretary of Commerce transmitting NOAA's comments. NOAA concurred with all five of our recommendations and identified steps that it is taking to implement them. It also provided technical comments, which we have incorporated into our report, as appropriate. In its technical comments, NOAA suggested we remove our discussion of a programmatic risk associated with a contractor's sale of its server business to a foreign company. NOAA's CIO stated that the terms of the sale ensured that this risk was addressed. We removed this risk from our report, but will continue to monitor the agencies' efforts to manage and oversee the maintenance of its servers.

On October 17, 2014 an audit liaison for the Department of Defense provided an email stating that the department did not have comments on the report's findings or recommendations. The liaison also provided technical comments, which we have incorporated into our report.

On October 22, 2014 an audit liaison for NASA provided an email stating that the agency would provide any input it might have to NOAA for inclusion in NOAA's comments.

David A. Powner
Director, Information Technology Management Issues

APPENDIX I: OBJECTIVES, SCOPE, AND METHODOLOGY

Our objectives were to (1) evaluate NOAA's progress on the JPSS satellite program with respect to cost, schedule, and mitigation of key risks; (2) identify the benefits and challenges of alternatives for polar satellite gap mitigation; and (3) assess NOAA's efforts to establish and implement a comprehensive contingency plan for potential gaps in polar satellite data.

To evaluate NOAA's progress on the JPSS satellite program with respect to cost, schedule, and mitigation of key risks, we assessed to what extent the program was (1) meeting expected costs for key program components; (2) meeting expected schedules for key program milestones; and (3) effectively mitigating key program risks through implementation of action plans and assessing the potential for gaps in polar afternoon satellite data. A more detailed description of our activities in each of these areas follows.

- **Cost:** We reviewed documentation from the program's key milestone review in July 2013 and program management council briefings to identify baseline and estimated costs for key JPSS program components as of July 2013. We reviewed monthly program status reports through July 2014 and compared them to the program's prior cost estimates to identify any changes in estimated costs and to determine the reasons why development costs had increased or decreased. We assessed the reliability of the cost estimates by interviewing program officials to determine the process the program used to formulate the cost estimates and to conduct accuracy checks on data presented in program status reports. We determined that the cost data were sufficiently reliable for our reporting purposes. We also interviewed program officials to discuss JPSS development status and reasons for changes in costs.
- **Schedule:** We reviewed documentation from the program's key milestone review in July 2013, including the program's integrated master schedule developed for this review, to select program milestones representing key near-term development and identify baseline and estimated completion dates for those milestones as of July 2013. We reviewed documentation prepared for the JPSS-1 mission critical design review and monthly program status reports and compared them to expected completion dates to actual/estimated completion dates as of July 2014 for the selected milestones. We reviewed monthly program status reports to identify reasons why

milestone dates may have changed. We assessed the reliability of reported milestone dates by examining multiple program and project status reports at different points in time for consistent reporting of dates or explanations of any changes and compared reported dates to source schedule data. We determined that the milestone data were sufficiently reliable for our reporting purposes. We also interviewed agency officials to review our selection of milestones and to determine additional rationale for changes in milestone schedules.

- **Risk:** We reviewed documentation from the program's key milestone review in July 2013 to identify top program risks. We reviewed program monthly status reports through July 2014 to track the status of these and other top program risks presented monthly to the NOAA program management council. We reviewed monthly program status reports, documentation on risk mitigation action plans, risk management board meeting minutes, and JPSS project risk registers to assess (a) whether the program had developed a mitigation plan with specific actions to address the risk; (b) whether the program tracked the status of mitigation plan implementation; (c) the extent to which the mitigation plan was implemented on schedule; and (d) the extent to which the program reduced the risk's likelihood and consequence. We also evaluated whether the program tracked actual completion dates for mitigation steps against the planned schedule, as recommended by leading organizations.[1] We obtained an October 2013 briefing and detailed report with the JPSS program's assessment on the potential for a gap in polar satellite data and assessed key assumptions from this study. We compared the results of the study with prior GAO reports and briefings from NASA's Orbital Debris Program Office on future projections of orbital debris risk to determine scenarios that could result in polar satellite gaps. We interviewed the program's risk manager and other relevant program officials to discuss the program's risk mitigation processes, the status of risk mitigation action plans, and the program's assessment of potential gaps in the polar afternoon orbit.

To identify benefits and challenges of alternatives for polar satellite gap mitigation, we reviewed existing literature discussing alternatives for polar satellite gap mitigation and developed a list of alternatives that have been recently identified for NOAA to consider. We interviewed experts from 15 different organizations within and outside of NOAA—including NOAA organizations with key responsibilities for gap mitigation efforts, NOAA weather forecast offices, international weather forecast organizations, Department of Defense weather forecast organizations, industry associations and contractors, and research organizations related to weather forecasting and modeling—to gather input on benefits and challenges associated with the list of alternatives.[2] We categorized the list of alternatives based on the capabilities they were expected to provide and summarized the key benefits and challenges that experts commonly identified for the alternatives.

To assess NOAA's efforts to establish a comprehensive contingency plan for potential gaps in polar satellite data, we reviewed NOAA's February 2014 polar satellite gap contingency plan and compared it against contingency planning best practices as well as shortfalls we previously identified in NOAA's October 2012 contingency plan. We compared elements of the plan against best practices developed from leading government and industry sources such as the National Institute of Standards and Technology, the Software Engineering Institute's Capability Maturity Model® Integration, and our prior report. 3 In order to assess NOAA's efforts to implement a comprehensive contingency plan for potential gaps in polar satellite data, we reviewed NOAA's February 2014 contingency plan, progress reports for relevant projects, and relevant project plans and proposals to identify the set of projects NOAA has been implementing and the planned schedule and outcomes of those projects. We evaluated project status reports to determine the extent to which NOAA has reported on and tracked the progress of its polar gap mitigation projects and the extent to which NOAA has been meeting each project's planned schedule and outcomes. We also evaluated the extent to which NOAA had prioritized its mitigation projects by comparing them to best practices identified by the Office of Management and Budget and from prior GAO reports.[4] We interviewed relevant NOAA and JPSS program officials regarding changes reflected in the February 2014 contingency plan, oversight and reporting procedures for gap mitigation projects, and implementation progress for gap mitigation projects.

We performed our work at NASA and NOAA offices in the Miami, Florida, Boulder, Colorado, and Washington, D.C. area. We conducted this performance audit from January to December 2014 in accordance with generally accepted government auditing standards. Those standards require that we plan and perform the audit to obtain sufficient, appropriate evidence to provide a reasonable basis for our findings and conclusions based on our audit objectives. We believe that the evidence obtained provides a reasonable basis for our findings and conclusions based on our audit objectives.

Appendix II: Space Debris Pose a Catastrophic Risk to Spacecraft

Experts estimate that more than 100 million man-made objects larger than 1 millimeter are orbiting the earth. These objects—space debris— include whole and fragmentary parts of rocket bodies and other discarded equipment from space missions. According to NASA's Orbital Debris Program Office, over 500,000 of these objects (those 1 centimeter and larger) pose a threat to current and future satellites because they can cause catastrophic damage upon impact.[1]

NASA and DOD share responsibility for determining the amount of space debris at various sizes and altitudes. DOD uses its Space Surveillance Network, which is comprised of radars and electro-optical sensors around the world, to track and catalog objects as small as 5 centimeters in diameter. While smaller debris cannot be tracked, NASA's Orbital Debris Program Office estimates the amount of smaller debris using ground-based radars and visual inspections of spacecraft that have returned from orbit. It also develops and maintains models of the space debris environment at different altitudes. NASA currently estimates that in low-earth orbit the debris populations are approximately 22,000 objects larger than 10 centimeters; approximately 500,000 objects between 1 and 10 centimeters in diameter; and roughly 100 million particles between 1 millimeter and 1 centimeter.

Due to the high impact speed in space (approximately 10 kilometers per second, or 22,369 miles per hour), even debris that is less than 1 centimeter can cause catastrophic damage to a spacecraft.[2] As an example, space debris as small as 0.3 centimeters could have the same effect as a bowling ball traveling at 60 miles per hour. As debris size increases, so does the potential impact to spacecraft. For example, debris as small as 1 centimeter can have the same effect as a 400 pound safe traveling at 60 miles per hour, debris that is 5 centimeters has energy similar to being hit by a bus, and debris that is 10 centimeters has energy similar to a large bomb. The following figure illustrates the potential effects of various sizes of space debris.

Options for mitigating the impacts of space debris differ depending on the size of the debris. For debris 5 to 10 centimeters and larger that are typically tracked and cataloged, spacecraft operators can perform collision avoidance maneuvers. For debris that is less than 2 centimeters, shielding can generally provide protection to critical components. However, there are limitations to adding shielding, such as the amount of weight that it adds to the spacecraft and the inability of shielding to protect from debris larger than 2 centimeters.[3] Because debris less than 5 centimeters is not tracked and protective shielding is generally limited to debris 2 centimeters and smaller, there is no mitigation for debris in this size range other than reorienting the spacecraft or flying the spacecraft at altitudes with less debris. However, this may not be an option for certain missions like JPSS that requires a specific spacecraft orientation in a polar sun-synchronous orbit at an altitude of 824 kilometers.

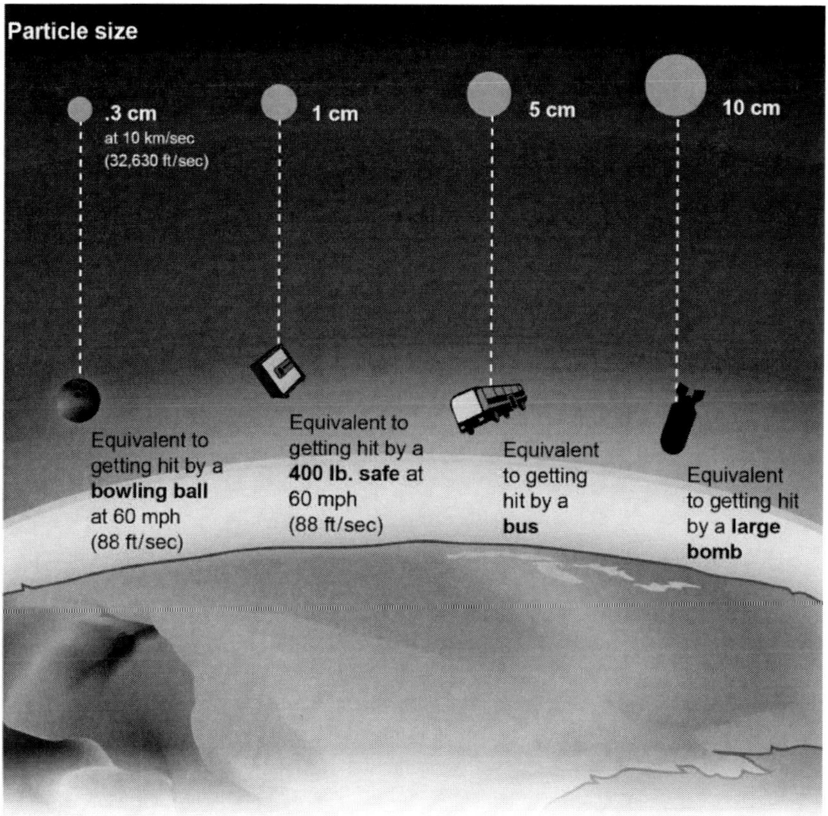

Source: Report on Orbital Debris, by Interagency Group (Space), National Security Council, February 1989, and data used with permission of The Aerospace Corporation. | GAO-15-47.

Figure 6. Potential Effects of Space Debris at Various Sizes.

JPSS Satellites' Orbit Has an Elevated Risk from Space Debris

Space debris experts from NASA's Orbital Debris Program Office believe that the danger of collisions to satellites at altitudes similar to JPSS is compounded by estimates that the amount of space debris will increase as worldwide spacecraft launch rates increase and collisions between debris particles create more debris. In particular, experts are concerned about increases in space debris that is larger than 2 centimeters for which shielding options are limited and less than 5 centimeters for which collision avoidance maneuvers cannot be conducted because the debris are not actively tracked and cataloged. Figure 7 illustrates NASA's projection of the amount of space debris that is 0.3 centimeters and larger at JPSS satellites' altitude through 2035.

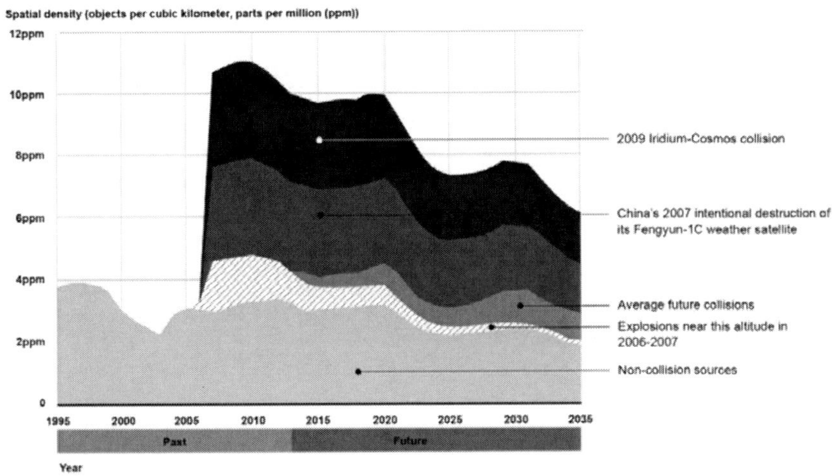

Source: NASA Orbital Debris Program Office. | GAO-15-47.

Note: Spatial density represents the effective number of spacecraft and other objects as a function of altitude. In this figure, spatial density represents the number of objects larger than 0.3 centimeters at an altitude of 825 kilometers.

Figure 7. Projected Space Debris at JPSS Satellites' Altitude.

End Notes

[1] NOAA, NASA, and the Department of Defense were jointly responsible for the National Polar-orbiting Operational Environmental Satellite System (NPOESS). While the disbanding of the program began in 2010, NPOESS was officially terminated in 2011.

[2] GAO, *High Risk Series: An Update*, GAO-13-283 (Washington, D.C.: February 2013).

[3] GAO, *Polar Weather Satellites: NOAA Identified Ways to Mitigate Data Gaps, But Contingency Plans and Schedules Require Further Attention*, GAO-13-676 (Washington, D.C.: Sept. 11, 2013).

[4] According to NOAA, 80 percent of the data assimilated into its National Weather Service numerical weather prediction models that are used to produce weather forecasts three days and beyond comes from polar-orbiting satellites.

[5] The European Organisation for the Exploitation of Meteorological Satellites' Metop program is a series of three polar-orbiting satellites dedicated to operational meteorology. Metop satellites are planned to be flown sequentially over 14 years. The first of these satellites was launched in 2006, the second was launched in 2012, and the final satellite in the series is expected to launch in 2017.

[6] Presidential Decision Directive NSTC-2, "Convergence of U.S. Polar-orbiting Operational Environmental Satellite Systems," May 5, 1994.

[7] Originally called the NPOESS Preparatory Project, in January 2012 the satellite's name was changed to the Suomi National Polar-orbiting Partnership satellite.

[8] GAO, *Polar-orbiting Operational Environmental Satellites: Technical Problems, Cost Increases, and Schedule Delays Trigger Need for Difficult Trade-off Decisions*, GAO-06-249T (Washington, D.C.: Nov. 16, 2005).

[9] Using S-NPP as an operational satellite means that the satellite's data will be used to provide climate and weather products.

[10] The announcement accompanied the release of the President's fiscal year 2011 budget request.

[11] This figure does not include approximately $2.9 billion in sunk costs that DOD spent on NPOESS through fiscal year 2010.

[12] NOAA's $3.3 billion sunk costs included $2.9 billion through fiscal year 2010 and about $400 million in fiscal year 2011.

[13] SIDAR is to accommodate the Total and Spectral Solar Irradiance Sensor, the Advanced Data Collection System, and the Search and Rescue Satellite-Aided Tracking system.

[14] This practice is not consistent with the common definition of a life cycle cost estimate. See *GAO Cost Estimating and Assessment Guide: Best Practices for Developing and Managing Capital Program Costs,* GAO-09-3SP (Washington, D.C., March 2009).

[15] See, for example, GAO, *Polar-orbiting Environmental Satellites: Changing Requirements, Technical issues, and Looming Data Gaps Require Focused Attention*, GAO-12-604 (Washington, D.C., June 15, 2012); *High Risk Series: An Update*, GAO-13-283 (Washington, D.C.: February 2013); and GAO-13-676.

[16] The European Centre for Medium Range Weather Forecasts is an independent, intergovernmental organization supported by 34 European nations, providing global medium-to-extended range forecasts.

[17] GAO-12-604.

[18] GAO-13-283.

[19] GAO-13-676.

[20] The critical path is generally defined as the longest continuous sequence of activities in a schedule. As such, it defines the program's earliest completion date or minimum duration. If

an activity on the critical path is delayed by a week, the program finish date will be delayed by a week unless the slip is successfully mitigated. Therefore, the critical path is most useful as a tool to help determine which activities deserve focus and, potentially, management help.

[21] See, for example, Software Engineering Institute, *CMMI® for Acquisition, Version 1.3* (Pittsburgh, Pa.: November 2010) and Project Management Institute, *A Guide to the Project Management Body of Knowledge* (PMBOK® Guide), Fifth Edition, (Newton Square, Pa.: 2013). "PMBOK" is a trademark of the Project Management Institute, Inc.

[22] NOAA officials noted that while it has taken more than 2 years to fully validate certain high priority products for S-NPP, data from ATMS were used operationally 7 months after the satellite's launch. Further, NOAA officials believe that data from instruments with a history of operational use can occur within 4 months of launch if they have similar data quality.

[23] Software Engineering Institute, *CMMI® for Acquisition, Version 1.3.*

[24] See GAO, *Year 2000 Computing Crisis: Business Continuity and Contingency Planning*, GAO/AIMD-10.1.19 (Washington, D.C.: August 1998); National Institute of Standards and Technology, *Contingency Planning Guide for Federal Information Systems*, NIST 800-34 (May 2010); Software Engineering Institute, *CMMI® for Acquisition, Version 1.3* (Pittsburgh, Pa.: November 2010).

[25] GAO-13-676.

[26] Office of Management and Budget, *Capital Programming Guide: Supplement to Circular A-11, Part 7, Preparation, Submission, and Execution of the Budget* (Washington, D.C.: Executive Office of the President, July 2014).

[27] See GAO, *2020 Census: Prioritized Information Technology Research and Testing is Needed for Census Design Decisions*, GAO-14-389 (Washington, D.C.: April 3, 2014); and GAO-12-120G.

End Notes for Appendix I

[1] See, for example, Software Engineering Institute, *CMMI® for Acquisition, Version 1.3* (Pittsburgh, Pa.: November 2010) and Project Management Institute, *A Guide to the Project Management Body of Knowledge* (PMBOK® Guide), Fifth Edition, (Newton Square, Pa: 2013). "PMBOK" is a trademark of the Project Management Institute, Inc.

[2] The NOAA organizations include the National Centers for Environmental Prediction's Environmental Modeling Center and National Hurricane Center; the National Weather Service's Alaska Weather Forecast Offices (Anchorage and Fairbanks); the Office of Atmospheric Research's Atlantic Oceanographic and Meteorological Laboratory and Earth System Research Laboratory; the Office of the Chief Information Officer; and the National Environmental, Satellite, Data and Information Service's Integrated Product Team (which includes representatives from the JPSS Program Office and Office of Satellite and Product Operations). The international weather organizations include the United Kingdom Meteorological Office and the European Centre for Medium-range Weather Forecasts. The DOD entities include the Air Force Weather Agency and the Fleet Numerical Meteorology and Oceanography Center. The industry and contracting organizations include the University Corporation for Atmospheric Research and the Alliance for Earth Observations. The research organizations include NASA's Global Modeling and Assimilation Office and Jet Propulsion Laboratory.

[3] GAO-13-676.

[4] Office of Management and Budget, *Capital Programming Guide: Supplement to Circular A-11, Part 7, Preparation, Submission, and Execution of the Budget* (Washington, D.C.: Executive Office of the President, July 2014); GAO, *2020 Census: Prioritized Information Technology Research and Testing is Needed for Census Design Decisions*, GAO-14-389 (Washington, D.C.: April 3, 2014); and GAO-12-120G.

End Notes for Appendix II

[1] According to a NASA expert who conducts hypervelocity impact testing, space debris less than 1 centimeter can cause serious damage to satellites, possibly resulting in mission termination. The impacting particle size causing failure is a function of satellite design and shielding. Particle sizes from 1 to 5 millimeters typically cause failure of important satellite hardware such as propellant tanks.

[2] GAO, *Space Program: Space Debris a Potential Threat to Space Station and Shuttle*, GAO/IMTEC-90-18 (Washington, D.C.: April 6, 1990).

[3] According to the program manager of NASA's Orbital Debris Program Office, most robotic spacecraft are not shielded against debris as large as 1 centimeter because shielding for debris of this size requires both weight and volume that is typically not available.

INDEX

A

B

C

O

P

Q

R

V

W